"Gina Says"

Adventures in the
Blogosphere String War

"Gina Says"

Adventures in the Blogosphere String War

Gil Kalai

Hebrew University of Jerusalem, Israel

World Scientific

NEW JERSEY · LONDON · SINGAPORE · BEIJING · SHANGHAI · HONG KONG · TAIPEI · CHENNAI · TOKYO

Published by

World Scientific Publishing Co. Pte. Ltd.
5 Toh Tuck Link, Singapore 596224
USA office: 27 Warren Street, Suite 401-402, Hackensack, NJ 07601
UK office: 57 Shelton Street, Covent Garden, London WC2H 9HE

Library of Congress Cataloging-in-Publication Data
Names: Kalai, Gil, 1955–   author.
Title: "Gina says" : adventures in the blogosphere string war / by Gil Kalai
    (Hebrew University of Jerusalem, Israel).
Description: Singapore ; Hackensack, NJ : World Scientific, [2017]
Identifiers: LCCN 2016054699| ISBN 9789813142060 (hardcover ; alk. paper) |
    ISBN 9813142065 (hardcover ; alk. paper) | ISBN 9789813142077 (pbk. ; alk. paper) |
    ISBN 9813142073 (pbk. ; alk. paper)
Subjects: LCSH: Science--Blogs. | Mathematics--Blogs. | String models--Blogs. | Physics--Methodology--Blogs.
Classification: LCC QC6 .K245 2017 | DDC 539.7/258--dc23
LC record available at https://lccn.loc.gov/2016054699

British Library Cataloguing-in-Publication Data
A catalogue record for this book is available from the British Library.

Typeset by Stallion Press
Email: enquiries@stallionpress.com

To my family

Debates portrayed in books, are the worst sort of readings

Jonathan Swift

Preface

In the summer of 2006 two books attacking string theory, a prominent theory in physics, appeared: One by Peter Woit called "**Not even wrong**" and the other by Lee Smolin called "**The trouble with Physics**." A fierce public debate, much of it on weblogs, ensued.

Gina is very curious about science blogs. Can they be useful for learning about, or discussing science? What happens in these blogs and who participates in them? Gina is eager to delve into the issues at hand and to form her own opinion about the string theory controversy. She is equipped with some relevant academic background, and has some familiarity with academic life. Her knowledge of physics is derived mainly from popular accounts. Gina likes to debate and to argue and to be carried by her associations. She is fascinated by questions about rationality and philosophy, and was exposed to various other scientific controversies in the past.

This book uses the blog string theory debate to talk about blogs, science, and mathematics. Meandering over various topics from children's dyscalculia to Chomskian linguistics, the reader may get some sense of the chaotic and often confused scientific experience. The book tries to show the immense difficulty involved in getting the factual matters right and interpreting fragmented and partial information.

Gina's writings on the blogosphere were selected and edited by Gil Kalai, Professor of Mathematics and member of the Center for the Study of Rationality at the Hebrew University of Jerusalem and Professor of Computer Science and Mathematics at Yale University. (The original blog postings were often shortened.)[1]

Short academic pieces associated with the blog discussions were written with the help of colleagues (and Wikipedia), and are interspersed throughout the book along with an appendix on some puzzles in mathematics. The editor would like to thank Anat Amirav, who edited the English and contributed two chapters, Michal Linial, who contributed a chapter, and his daughter, Neta Kalai, for her beautiful drawings.

---

[1]The full blog discussion is available on the blogs: Not Even Wrong http://www.math.columbia.edu/~woit/wordpress/; n-Category Cafe, https://golem.ph.utexas.edu/category; Asymptotia http://asymptotia.com/.

Contents

Prologue: String Theory and the Debate Surrounding It

What is string theory? String theory is a very ambitious (yet incomplete) answer to the problem, prevalent on the minds of physicists for almost a century, of bringing together quantum theory and Einstein's theory of gravitation. String theory is delicate enough to include (at least in principle) the successful twentieth-century physics theories of elementary particles. (The "Standard Model" refers to the description given by these successful theories for the three out of four fundamental forces in nature, and for the particles associated with these forces.) Although string theory is far from complete, some of its insights are surprisingly detailed and robust. String theory with its many startling contributions within physics and to mathematics is widely considered to be humanity's best shot at understanding the most fundamental questions about our physical reality.

Is it possible that string theory, a major physics theory described as a crowning intellectual human achievement, is simply false, and the efforts of thousands of brilliant researchers for more than two decades have been misguided? Even worse, is it the case that string theory fails even to lead to any predictions and cannot be regarded as a scientific theory? Has this failure been covered up by prominent scientists who simply ignore the evidence in front of them, use their stature to promote failed ideas, and artificially hype them while basically refusing to admit failure? Are these scientists ready to modify the definition of science, scientific standards, and the ethics of science, just to have it their way? And could it be that they not only failed, but even led to the fall of science as a whole? Such claims against string theory have been heard on occasion in the last two decades. The books by Peter Woit and Lee Smolin that appeared in the summer of 2006 put together various arguments against string theory as well as against the social aspects of academic life and universities, and ignited a fierce public debate.

This book is not about string theory. String theory and the debate around it are the envelope for the events and issues described in this book. It may be useful for the reader, now or while reading the book, to know more about the major insights of string theory and for that the reader may go on reading our very short description of string theory and may also use the small dictionary following it. Otherwise, the reader can proceed directly to the debate on page 1, even knowing as little as Gina did when she got into it.

String theory — a very short description

Some of the major insights of string theory can be described as follows:

1) Point particles are replaced by strings, one-dimensional geometric objects. In later variants of the theory, more general geometric objects called branes also play a role.

2) These strings exist in a high-dimensional space. In other words, aside from the familiar three dimensions of our world and the one dimension needed to describe time, there are six (or more) additional tiny dimensions.

While these two basic insights of string theory are already quite confusing and may look rather strange, the next three insights use jargon which we ask the reader to take for granted. Detailed popular explanations can be found in various books and articles, like Brian Greene's bestseller *The Elegant Universe*. (We will return to "high dimensions" and some of the other technical notions later in the book.)

3) The additional six dimensions are described by mathematical objects called "Calabi-Yau manifolds."

4) String theory is based on "supersymmetry," a new form of symmetry for the laws of physics.

5) String theory also offers "dualities," which is a name for deep connections between different string theories. Dualities discovered by string theorists revolutionized some areas of mathematics.

6) Another important insight of string theory is that of a mysterious "M-theory," which unifies various models for string theory into a single entity.

7) Another more recent insight is the discovery that there is an extended (huge) family of different universes described by string theory. This phenomenon is referred to as the "landscape of vacua of string theory." This discovery has led to

various foundational questions regarding string theory as a scientific theory, both among string theorists and among its critics.

All these insights are, to some extent, "negotiable." There are variations of string theory, where a few of these ingredients are omitted and replaced by others.

The technical study of string theory is, of course, not based on these verbal descriptions (nor even on the far more detailed descriptions that can be found in several popular books), but on complicated mathematical formulations of these and related ideas. Some of the critique on string theory is very technical, but much of it, and certainly the points that appeal to a wider audience, is nontechnical and refers to general issues regarding the philosophy of science, and especially the need for empirical validations of scientific theories, the relations between physics and mathematics, and the sociology of scientific life.

The history of string theory and the stories of scientists involved in it is a fascinating story all on its own, and one which we will not go into here. I will mention one name only, Edward (Ed) Witten of Princeton, one of the most decorated living scientists both in physics as well as in mathematics, who has been a dominant figure in string theory for many decades, and naturally he is often mentioned in the debate regarding string theory.

Following is a small dictionary of terms about physics and string theory.

A Small Dictionary

With the kind help of Wikipedia, I have compiled a short dictionary of some recurring "buzz words" in the string theory discussion. (Alas, sometimes the definition of a term sounds as cryptic as the term itself.) *To the reader: skip this part and return to it only if needed.*

Quantum electrodynamics (QED) — The classical theory of electricity and magnetism is among the crowning achievements of nineteenth-century physics and associated mainly with the work of Maxwell. Quantum electrodynamics describes electricity and magnetism in the framework of quantum mechanics and, in particular, the mathematical rules according to which electrons, positrons, and photons interact. The theory was developed over the first half of the twentieth-century, and is considered the "jewel of physics" for its extremely accurate predictions.

Quantum chromodynamics (QCD) — The theory of the strong interaction, a fundamental force describing the interactions of "quarks," which are the building blocks for particles like the proton and the neutron.

The Standard Model — A theory that describes three of the four known fundamental interactions between the elementary particles that make up all matter. (QED and QCD are parts of the Standard Model.) The missing force not described by the Standard Model is gravitation.

Quantum gravity — A theory that unifies quantum physics and the theory of gravitation.

String theory — A prominent candidate for the theory of quantum gravity that includes the Standard Model. The main idea of string theory is to replace point-particles by one-dimensional geometric objects called "strings." A bold consequence of string theory is that the universe has extra dimensions. The first version of the theory studied a 26-dimensional universe and more recent versions (based on "supersymmetry") assert that the universe has 10 or 11 dimensions.

Supersymmetry (SUSY) — A new kind of symmetry studied in theoretical physics and mathematics, and yet to be confirmed empirically as a rule of physics. Supersymmetry suggests that every known elementary particle has a yet-to-be-discovered "super partner," and offers a possible explanation of "dark matter," which accounts for most of the mass in the universe. Supersymmetry also plays an important role in string theory.

Loop quantum gravity (LQG) — An approach to unifying gravitation and quantum mechanics that differs from that of string theory.

The landscape — String theory seems to allow an astronomical number of different possible universes. (The number is taken to be 10^{500}, that is, one followed by 500 zeroes.)

The anthropic principle — A controversial approach to studying the rules of physics according to which humans should take into account the constraints on the rules of physics that human existence imposes. The anthropic principle is offered as a way of dealing with the landscape.

Particle physics phenomenology — The part of theoretical particle physics that deals with the application of the theory to high-energy particle physics experiments. Within the Standard Model, phenomenology is the calculation of detailed predictions for experiments, usually at high precision. Beyond the Standard Model, phenomenology addresses the experimental consequences of new models.

The Large Hadron Collider (LHC) — A particle accelerator and collider located near (or underneath) the French-Swiss border. After many years of constructions the LHC began operation in September 2008. The LHC is the world's largest and highest-energy particle accelerator. It is hoped that the LHC will confirm some predictions of the Standard Model and will allow for the testing of further insights of theoretical physics, like supersymmetry.

Part I

NOT EVEN WRONG:
The Blog of Peter Woit

1 Extraordinary Claims

Virtually all of string theorists are nice people who never argue with anyone else, they're not chauvinists, and most of them are feminists.

Lubos Motl, August 30, 2006

Extraordinary claims require extraordinary evidence

Gina Says:
August 30th, 2006 at 7:39 am

Peter, is it possible to state the main points for the "case against string theory" with 4–5 sentences on each? This will be very helpful. Please consider doing it. It will be useful to

a) Separate the strictly scientific points from more sociological and philosophical points

b) Separate points that say that (*) "string theory is not (yet) successful" from those saying that (**) "this and that aspects of the theory are fishy" from those saying that (***) "string theory is not the right direction for dealing with 'final theory/grand unification'", from those saying that (****) "the whole endeavor of final theory/grand unification is misguided."

c) Hint whether physics requires more back-tracking or whether it is just string theory that is problematic.

Thanks!

Gina's first comments generated several answers: 'a' said that string theory's main scientific problems are two-fold: "first, since we cannot directly probe quantum gravity, a useful theory of quantum gravity must predict something at the lower energies where we can conduct experiments, and second, that string theory allows something like 10^{500} different possible theories and that this seems practically equivalent to allowing everything and predicting nothing."

Peter Woit Says:
August 30th, 2006 at 1:36 pm

Gina,

At this point, I'm way too busy, and writing another explanation of what the problems are with string theory isn't at all something I want to spend time on. The article I wrote back in 2001 is still a good short version of the argument; all it is missing is a discussion of how things have gotten much, much worse for string theory since then, because of the landscape.

The issues involved here are pretty complicated, and I don't think short sound-bites, or me countering people's "10 quick reasons why string theory is great" with "10 quick reasons why string theory doesn't work" is going to be very enlightening. Partisans of one point of view or the other aren't going to be convinced by this, and people who want to seriously understand the issues and make up their own minds should read both the pro-string theory point of view put forward in several books, and the other side of the story, as explained in my book and in Lee's. I do believe that the problem is not just string theory, but more generally the idea of supersymmetric grand unification; these issues are discussed extensively in the book.

Gina was happy with Peter's answer. Especially interesting was his skeptical point of view, not just toward string theory, but also toward the idea of "supersymmetry." Another answer came from Nigel.

Nigel Says:
August 30th, 2006 at 8:38 am

Gina,

Your idea would, I fear, produce a list of string theory claims with the same boring label 'uncheckable speculation' beside each.

'Extraordinary claims require extraordinary evidence.' — Carl Sagan.

Gina fondly remembered what her great uncle Lena used to say about extraordinary claims:

"Why should I be surprised if I can simply disbelieve."

As for Sagan's line on extraordinary claims, Gina later discovered that quoting it is quite popular on both sides of this debate, and other debates as well. Gina found Sagan's statement rather confusing. She felt that, just like ordinary claims, extraordinary claims must be based on good evidence, and that the problem with the many extraordinary claims that Sagan talked about (flying saucers, astrology), was that the evidence was simply lacking. Sagan's statement led to endless discussions about which side of the debate is extraordinary to begin with, and it was not clear what "extraordinary evidence" really meant.

Nigel continued:

> You can see plenty of extraordinary claims in string theory (it solves almost all the big problems of unification, quantum gravity, the nature of particles, black holes). You don't see any stringy evidence, let alone extraordinary evidence, and nobody expects to find much.
>
> The string theory failure has some weak precedents in science: the "Vortex Atom" and "Aether" (both the subjects of intricate mathematical speculation and wild claims of ad hoc success from mathematical physicists including Kelvin and Maxwell, who both died firmly believing flawed theories).
>
> However, string theory is more dangerous. At least Kelvin and Maxwell's ideas could later be checked by experiment. String theory deliberately speculates about practically uncheckable phenomena (Planck scale unification, etc) thus remaining safe from experimental refutation, and consequently string theory is becoming a religion:
>
> 'Whatever ceases to ascend, fails to preserve itself and enters upon its inevitable path of decay. It decays ... by reason of the failure of the new forms to fertilize the perceptive achievements which constitute its past history.' — Alfred North Whitehead, F.R.S., Sc.D., Religion in the Making, Cambridge University Press, 1927, p. 144.

Gina was happy with the responses. She was unclear as to why dying with a firm belief in a flawed theory was worse than simply dying, and how string theory could possibly be dangerous, but found it completely convincing and yet too general to be relevant, that what stops to ascend may enter a path of decay. So she asked:

Gina Says:
August 30th, 2006 at 2:10 pm

Is the critique of string theory similar to the critique on the biologists for not understanding/finding a cure for cancer? Or is it stronger?

I think the main reason for me to be suspicious about "string-theory bash-ing" is that it does not lead to interesting science: namely to scientific papers (not popular reviews and books). Why is that?

Dying with a firm belief in a flawed theory

Peter Woit once more gave a thoughtful answer: He regarded the situation as far worse than cancer research and added: "an analog would be if current cancer treatments not only didn't help at all with the disease, but made it much worse." He explained that "string-theory bashing" does not lead by itself to new science, but encourages people to leave "failed ideas" behind and look for something else. And he pointed out that there are alternatives to string theory and that he has his own ideas regarding alternatives.

Another key figure in the debate on Woit's blog and on his own blog, was Lubos Motl, a Harvard physicist. Expressing himself quite bluntly, Lubos referred to Woit and Smolin as crackpots, and as the enemies of science. Some of his opinions came across as chauvinistic and racist. When Gina conjectured that Lubos' attitude was highly uncharacteristic of string theorists, it was Lubos himself who quickly supported her conjecture.

Lubos Motl Says:
August 30th, 2006 at 6:47 pm

Virtually all of string theorists are nice people who never argue with anyone else, they're not chauvinists, and most of them are feminists. Most of them also think that string/M-theory are robust twin towers that are not threatened by any social effect or passionate proponents of alternative theories or proponents of no theories, and they almost always try to avoid interactions that could lead to tension, which also gives them more time for serious work. Almost no string theorists drive SUVs and they produce a minimum amount of carbon dioxide.

Why should I be surprised if I can simply disbelieve

Gina was happy to partake in the conversation and felt she got it going. She was fascinated with the other bloggers and she was curious as to who they were. Some of them had strange names and she could not understand the reason for picking such names.

The discussion received an interesting twist when a blogger mentioned logic and the famous work of Gödel's theorem as an analogy to the case at hand. We will come to this in the next chapter, but first here is a brief description of some of the participants in the "blogosphere string war."

Who are the Bloggers?

Peter Woit (left) and Lee Smolin (right)

Roughly two thirds of the comments on the weblogs that Gina participates in were made by anonymous bloggers. Some of these anonymous entities have been taking part in blog discussions for years. A few prominent anonymous participants are 'Nigel' (Chapter 1), 'Ebgert' (Chapter 2) who draws the analogy between string theory and Gödel's theorem, and 'The Graduate' (Chapter 3) who takes part in many discussions and who seems open minded and genuinely interested in learning. Here are a few more: The 'geometer' (Chapter 7) is apparently a mathematician based in geometry, and he often has interesting comments to make about geometry and other geometers. 'Renormalized' (Chapters 4 and 12) holds rather strong anti-string theory views and is, at times, quite hostile to Gina. 'Yatima' (Chapter 6) seems to have broad academic interests and knowledge, which are perhaps not rooted in the exact sciences. CapitalistImperialistPig (CIP, for short, Chapter 11) is quite an active participant and has a blog of his own. There are many more participants — in one heated debate (Chapter 11), there is one 'Anon' who sides with Gina, while yet another 'anon' takes the other side. Among the regular anonymous participants on the string theory side we can find 'moveon', 'amused' and 'aha'. One participant who contributes thoughtful remarks from time to time calls himself 'garbage'.

Among the bloggers who sign their full name, you can see, shoulder to shoulder, well known scientists, people outside academia with alternative physics theories of their own, and a wide spectrum of people in between. Quite a few of these participants also have blogs. For example, string theorist Jacques Distler (Chapter 32) has a blog called 'Musing,' Lubos Motl (Chapter 1), a string theorist who aggressively attacked Woit and Smolin, has a popular blog 'The reference Frame', 'bee' is the nickname of physicist Sabine Hossenfelder, who runs a blog called Backreaction, and 'Christine' (Chapter 34) is physicist Christine Dantas who ran a blog "Christine's Background Independence."

Mark Srednicki's participation in blog discussions regarding string theory is of special interest. Mark Srednicki was the chairman of the Department of Physics at the University of California at Santa Barbara. Srednicki works mainly in high energy physics. While making some contributions to string theory he is not a string theorist. Using his full name and academic credentials, Srednicki tried at a very early stage to shoot down Peter Woit's anti-string theory campaign, by commenting in Peter's blog. Later, he took part in a few discussions. In one comment, Srednicki shared interesting personal stories about the early stages of string theory and about his reactions to Witten's earlier discoveries. In a few other comments, he gave his point of view on various scientific issues that came up in the discussion. While patiently arguing with Lee Smolin and Peter Woit, it seems that his more recent approach towards the blog discussion and the entire debate was playful and relaxed, and he made a sincere effort to explain physics to laymen, in general, and his own views on matters at hand, in particular.

2 Gödel's Theorem and Logic

"…we conclude because A resembles B in one or more properties, that it does so in a certain other property."

John Stuart Mill, *"System of Logic Ratiocinative and Inductive* [1843]", Chapter XX on analogies.

A few days later, a blogger named 'Ebgert' attacked string theory, citing as an example Gödel's famous theorem about the incompleteness of mathematics. He drew the analogy between the landscape in string theory and Gödel's theorem in mathematics. Gina liked analogies and was curious about the value of analogies and about the way they were used in debates.

Gina Says:
September 3rd, 2006 at 5:51 am

Ebgert,

Your Gödel's theorem example goes the opposite way. There was a "foundational crisis" in mathematics based on the fear that its foundations were not provably sound. Gödel's theorem confirmed this fear, but strangely this was the end of the "crisis" and mathematics continued as before. There were people like Brouwer that thought and taught that mathematics should be done completely differently in view of these problems, but his views did not prevail. So if "landscape" to "string theory" is like "Gödel's theorem" to "mathematics", you can expect a bright future for string theory.

Ebgert Says:
September 3rd, 2006 at 6:43 am

Gina,

I agree with most of your points. In fact, Brouwer was correct, and mathematicians didn't pay as much attention to him as they should have.
Instead, they said "What's that weirdo talking about? Never mind, who
cares? Let's ignore him and call him a crackpot."

Ebgert complained about Physicists' "large egos" and continued:

"The physicists, string theorists in particular, will continue to insist that
they are talking about the ultimate truth?"

Gina was puzzled by Ebgert's last statement:

"My statement stands. Humans have *not* evolved significantly since the
McCarthy era; if you think otherwise you are wrong."

Gina Says:
September 3rd, 2006 at 7:29 am

Ebgert,

Hmmm, I am glad that you agree with most of my points. You come
across as being rather hostile towards physicists and mathematicians.
This seems unfortunate to me. Overall, it is not difficult to create public
hostility against intellectuals, scientists and various other minority and/
or "elite" groups. But I do not think this is a very good path to follow and
I doubt this is what Peter or Lee intend to do in their critiques. Brouwer,
right or wrong, was also a mathematician and he probably also had an
"ego" (a term coined by yet another intellectual).

Gina continued to mingle with the other bloggers. She got a feel for things
and she even read Peter Woit's 2001 article. To her surprise she could
understand it. The concern that string theory cannot be tested was prominent in this paper. Parts of the paper dealt with matters of public relations,

hype and funding, which Gina found interesting albeit not at the center of things. All in all, Gina was quite impressed by the many participants and by their knowledge. She was ready to look closer and more critically at some of Peter Woit's claims.

Gödel, Hilbert and Brouwer

Is mathematics a consistent theory? Or, rather, is there a danger of finding a correct mathematical proof for a false statement like "0 = 1"? These questions became quite relevant at the end of the nineteenth century, when some century-old mathematical truths were shattered and mathematicians started to feel the need for completely rigorous and solid foundations for their discipline.

Gödel's incompleteness theorem is a famous result in mathematics that shows the limitation of mathematics itself. At the end of the nineteenth century and the beginning of the twentieth, mathematicians tried to find a complete and consistent set of axioms for mathematics. This goal is often referred to as Hilbert's program, named after the mathematician David Hilbert who posed it as the second problem in his famous list of open problems in mathematics. In 1931 Kurt Gödel proved that this goal is impossible to achieve. Gödel proved that for any system of axioms for mathematics there are true results that cannot be proved! This is referred to as Gödel's first incompleteness result. One startling consequence is that it is impossible to precisely formulate the consistency of mathematics and therefore impossible to prove the consistency of mathematics. This is the content of Gödel's second incompleteness theorem.

Gödel's theorem is one of the few results of mathematics that capture the imagination of people well beyond mathematics. The well-known book *Gödel, Escher, Bach* by Douglas Hofstadter discusses common themes in the works of mathematician Gödel, artist M. C. Escher, and composer Johann Sebastian Bach.

Gödel's theorem is the climax (and, paradoxically, the end) of the "foundational crisis" in mathematics. Gottlob Frege made an important attempt to reduce all mathematics to a logical formulation. However, Bertrand Russell found a simple paradox that demonstrated a flaw in Frege's approach. The Dutch mathematician Luitzen E. J. Brouwer proposed an approach to mathematics, called intuitionism, which does not accept the law of excluded middle. This approach does not accept "Reductio ad absurdum", or, in other words, mathematical proofs "by contradiction". Most works in mathematics, including Brouwer's own famous earlier work, do not live up to the intuitionistic standards of mathematical proofs. Brouwer's ideas were regarded as revolutionary and, while on his lecture tours, he was received with an enthusiasm not usually associated with mathematics.

Hilbert and Brouwer were the main players in a famous controversy in the editorial board of *Mathematische Annalen*, the most famous mathematical journal of the time. Hilbert, the editor-in-chief, eventually fired Brouwer from the editorial board. There are different accounts regarding the nature of the disagreement. Some scholars have claimed that Brouwer wanted to impose his intuitionistic proof standards. Other scholars strongly reject this story and claim that Hilbert wanted to remove Brouwer in an inappropriate way simply because he felt that Brouwer was becoming too powerful.

Brouwer, Gödel, and Hilbert

3 Obviously I Disagree

"There's no success like failure, and failure's no success at all."

Bob Dylan, "Love Minus Zero/No Limit"

The title of Peter Woit's September 15th post was "Reviews in The Economist, Slate and the Times." Peter Woit discussed several reviews of his and Smolin's books, and Gina thought this would be a good opportunity to share her thoughts on Peter's book. A week earlier, Gina submitted a review, or as she called it a "pre-review," which was based on her initial impression of Woit's approach and book. She felt that Peter did not really have "a case" against string theory. However, it was posted in the wrong place and was deleted.

Gina Says:
September 16th, 2006 at 9:53 am

Let me remark how, in my opinion, the overall nice book by Peter (after all it is just a book, not a "case"), can be made nicer.

Most of the chapters of the book are quite good. I think this is a very good popularization of particle physics all the way to the "standard model". Popularization of science is a tricky business and deserves a whole separate discussion. There is no way to avoid some "cheating," but one should still try to be honest, useful and non-manipulative and Peter does a good job. The description of the connections with mathematics is especially interesting and the story about the Seiberg-Witten discovery is told very vividly.

With the exception of too strong rhetoric, most chapters on string theory are also well-written. I am learning a lot reading the book. Thank you, Peter!

What could make this book nicer? The story about the Bogdanov brothers and the reference to (the overplayed) Sokal's hoax do not contribute to the book. The same applies to the story about the string theory guy who became a Maharishi scientist, as well as to the description of the Cambridge University Press refereeing process. I have quite a few scientist friends, and complaining about referee reports is one of the few drawbacks in their sweet lives. Besides, they are referees themselves! (And they also complain about the burden of refereeing.) The rhetoric against string theory, and string theorists, as Peter himself noted (p. 225, l. -5) is indeed too strong. This does not add to, but rather reduces the value of the book.

The concluding chapter starts with a beautiful quote from Bob Dylan's song "Absolutely sweet Marie" –

"But to live outside the law, you must be honest".

When I first saw this quote I thought that this is a self-reference and that Peter set a standard for himself: If, coming from the outside, you want to single-handedly claim that one of the hottest scientific areas of our time failed, and that the efforts of thousands of string theorists are worthless, you'd better be honest about the details, presentation and even about your own motives. Apparently, Peter referred not to himself but rather to string theorists, for whom, in the absence of empirical support to their theory, honesty is paramount.

And Dylan's cryptic line from "Love Minus Zero/No Limit" also comes to mind:

"There's no success like failure, and failure's no success at all."

Based on a picture of Bob Dylan with Joan Baez (Washington D.C. 1963)

Debating string theory with quotations from Bob Dylan looked like the way to go!

Peter Woit Says:
September 16th, 2006 at 11:39 am

Gina,

Obviously I disagree with you about the relevance of some chapters in the book to my argument. In particular, the refereeing story at Cambridge was a very unusual one, involving two referees who strongly backed publication, and two string theory partisans trying (successfully) to stop Cambridge from publishing the book, while lacking any arguments against its content.

I should perhaps have made more explicit what I meant to convey with the Dylan quote. It's not specifically about me or about string theorists, but about the situation particle physics finds itself in. Lacking the discipline enforced by experiment, theorists now need to be a lot more self-critical and honest in evaluating the results of the speculative work they are engaged in.

Looking back later on her debate experience, Gina found it hard to think of a better phrase to describe the nature of blog debates than: **"obviously I disagree with you."** People hardly ever reversed or modified, or even reconsidered their opinions, not only on the big matters, but on the little points, as well.

Gina Says:
September 16th, 2006 at 4:29 pm

Peter: My main (mild) critique was not about irrelevancy, but rather about how the items I mentioned (and a few others), reduce the quality of your book. Of course, a book may have many qualities and I am mainly referring to the quality of the book as a serious discussion and debate of science. For example, if you refer to the string theory community as a "mafia", this statement is, of course, highly relevant, but making such a statement reduces the value of the book, at least in my opinion. I think you are wrong to consider your experience with Cambridge University Press as very unusual. Many authors had similar experiences, even with much less controversial (and more important) books and papers.

"The Graduate," another contributor expressed similar thoughts on the need for some "framing" in what Peter was writing. ("Who is 'the Graduate'?" thought Gina, remembering the movie and the famous Mrs. Robinson.)

Peter Woit Says:
September 16th, 2006 at 8:12 pm

Gina and the Graduate,

I did not refer to string theorists in the book as a "mafia", I said that some of the people who wrote to me did so. It is a fact that many people in the physics community feel this way and I was reporting this. It's not a word I would use to describe my own perception of string theorists' behavior. All the things that Gina objects to are things that I was reporting that are factually accurate.

Gina Says:
September 17th, 2006 at 1:49 pm

Peter,

The choices you have made make the book perfectly appealing for a commercial publisher but indeed not appropriate for a university press publication. Part of it is the rhetoric and selection of issues for discussion, and part of it is the clarity and strength of your overall argument.

Concerning the small issue of using the word "mafia": Indeed it is a quote from somebody else, but the discussion in the sentences which follow it gives the impression that you endorse what is behind this term, if not the term itself. This I find unfortunate.

In an extensive discussion in which there was much for Gina to learn, a discussion characterized by many misunderstandings, uncertainties and confusions — and, of course, fierce debate, it was comforting to be certain about something. And Gina was certain that Peter Woit's reference to string theorists as "mafia" was wrong.

A brief description of Peter Woit's book:

After a popular description of modern particle physics in Chapters 1–9, (which is easier for a mathematically inclined reader,) Chapter 11 goes on to describe string theory. Chapter 10 is devoted to a description of some startling insights from string theory to mathematics. The "case" against string theory is made in Chapters 12–18. Chapter 12 describes certain advantages offered by the notions of supersymmetry and string theory for settling fundamental problems in physics, as well as some difficulties associated with these notions which Woit regards as problematic. One such problem is the "hierarchy problem" which is a central puzzle in theoretical physics. The "landscape" is described and discussed in Chapter 17. The difficulty, perhaps even impossibility, of finding empirical evidence is emphasized in all chapters. In Chapter 13 on beauty and difficulty, Woit asserts that despite spending a lot of time learning string theory, he does not find it beautiful. He also regards the complexity of the theory as a negative indication. Chapter 14 raises the claim that string theory is not a science. A story about a string theorist who switched to studying transcendental meditation serves as a central example. Chapter 15 tells the story of two brothers, Igor and Grishka Bogdanov, who managed to publish several papers of no value in respectable physics journals. A famous hoax by Sokal, who intentionally submitted and published a worthless physics paper with a postmodernist flavor, is mentioned as an analogy. Chapter 16 tries to explain the dominance of string theory in spite of its alleged failure and the existence of some alternative approaches. Woit quotes an unnamed physicist referring to string theorists as "mafia", and further claims that a physicist criticizing string theory might need to worry about his or her professional safety! This chapter also describes Cambridge University Press's rejection of Woit's book.

Test your Intuition: What was Lord Kelvin's Main Mistake?

We now know that the age of the earth is $4.54 \pm 1\%$ billion years.

From Wikipedia: "In 1862, the physicist William Thomson (who later became Lord Kelvin) of Glasgow published calculations that fixed the age of Earth at between 20 million and 400 million years. He assumed that Earth had formed as a completely molten object, and determined the amount of time it would take for the near-surface to cool to its present temperature. His calculations did not account for heat produced via radioactive decay (a process then unknown to science) or convection inside the Earth, which allows more heat to escape from the interior to warm rocks near the surface."

Test your intuition:

What was the main reason for Lord Kelvin's wrong estimation?

a) Radioactivity — Heat produced by radioactive decay; this was a process which would remain unknown to science for decades to come.

b) Convection — The transfer of heat not through radiation or heat-conduction but through the movement of hot parts to the surface; this is a process common in home cooking.

4   "It is Not in the Cards"

The discussion continued. Usually, Gina's comments were premeditated, but from time to time she could not resist making a quick reply "on the fly". This time 'Dan' asked Peter for his prediction on the future of string theory.

Peter Woit Says:
September 17th, 2006 at 2:15 pm

My best guess is that string theorists will keep doing string theory no matter what, unless another bandwagon starts up for them to join. This might come about because of an exciting unexpected LHC result, because Witten comes up with a promising non-string theory idea, or some other reason. If the LHC doesn't find evidence for supersymmetry or extra dimensions (which I think is very likely), string theorists will concentrate more on black holes and cosmology (this has already been happening). While they do this, they'll slowly lose the support of their colleagues in other physics subfields, their funding will get cut, and, to a large extent, they'll take the whole field of theoretical particle physics slowly down with them.

Gina Says:
September 17th, 2006 at 4:07 pm

What you wrote, Peter, is an excellent generic guess for any prominent theory:

My best guess is that X-theorists will keep doing X-theory no matter what, unless another bandwagon starts up for them to join. This might come about because of an exciting unexpected EMPIRICAL result, because some prominent X-theorists come up with a promising non X-theory idea, or some other reason.

If no EMPIRICAL support will be found (which I think is very likely), X- theorists will concentrate more on possible application to theory Y (this has already been happening). As they do this, they'll slowly lose the support of their colleagues in other fields of their science, their funding will get cut, and, to a large extent, they'll take a large area of research slowly down with them.

Peter Woit Says:
September 17th, 2006 at 4:15 pm

Gina,

Yes, but in the generic case, a real possibility is that "X-theory will achieve one or more of its major goals, making it a solid and permanent part of science, opening up new areas to work on that build on this success." That's not at all in the cards in this case…

Gina Says:
September 17th, 2006 at 10:19 pm

"That's not at all in the cards in this case…"

Wow, so you have those magic cards, Peter! Boy, we have a lot of questions to ask you…

What's in the cards for strings?

Renormalized Says:
September 17th, 2006 at 5:03 pm

Gina- Do you always just regurgitate what others have written? I can't see you are adding anything to any discussion you have been involved with. Your responses are more in line with a common online troll.

This is uncalled for, thought Gina, it is legitimate to try to understand whether some of the critique against string theory actually has nothing to do specifically with string theory. Anyway, Gina's appeasing mood had changed. It was time to come back to her old review of Woit's book.

Thomas Bayes and Probabilities

"It is not in the cards" said Peter Woit. Later on, physicist Peter Orland referred to certain possible uses of probability reasoning in a discussion about an ambitious project, the string vacuum project. To quote Orland: "This probably won't happen, and isn't a proper application of probability anyway." Orland perhaps did not notice that the first part of his sentence also talks about probabilities in an improper way.

How can we assign probabilities in cases of uncertainty? And what is the nature of probabilities, to start with? And what is the rational mechanism for making a choice under uncertainty?

Thomas Bayes lived in the eighteenth century. Bayes' famous formula shows how to update probabilities given some new evidence. Following is an example for an application of Bayes' rule:

Suppose that ninety percent of pedestrians cross a certain crosswalk when the light is green, and ten percent cross it when the light is red. Suppose also that the probability of being hit by a car is 0.1% for a pedestrian who crosses on a green light, but the probability of being hit by a car is 2% for a pedestrian who crosses on a red light. A pedestrian is hit by a car at this particular crossing and brought to the hospital. How likely is it that he crossed on a red light?

Well, to start with (or *a priori*), only ten percent of the people who cross the crosswalk do so when the light is red, but now that we are told that this person was hit by a car it makes the probability that he crossed the street illegally higher. But by how much? Bayes' rule allows us to compute this (*a posteriori*) probability. I will not describe the mathematical formula, but I will tell you the outcome: the probability that this person crossed on a red light is 2/3.

The Bayesian approach can be described as follows. We start by assigning probabilities to certain events of interest and, as more evidence is gathered, we update these probabilities. This approach is applied to mundane decision-making and also to the evaluation of scientific claims and theories in the philosophy of science.

Bayes' rule tells us how to update probabilities, but we are left with the question of how to assign probabilities in cases of uncertainty to begin with. What is the probability of success in a medical operation? What is the chance of your team winning the next baseball game? How likely is it that war will break out in the Middle East in the next decade? What is the meaning of Orland's statement "this probably won't happen"? And what is the probability that string theory will prevail as the theory of quantum gravity?

One very early approach to probabilities, the principle of indifference (a.k.a. the principle of insufficient reason), asserts that given a certain number of mutually exclusive events, their probabilities are the same. The formulation of this principle goes back to Jakob Bernoulli and Pierre-Simon Laplace. This principle constitutes an important and very early appearance of the notion of **symmetry**. Of course, there are many cases where the principle of indifference fails miserably. Various other approaches to "subjective probabilities" and to the foundation of probability theory were developed in the twentieth century.

Decisions under uncertainty depend not only on the probabilities but also on the "stakes." Crossing a crosswalk on a red light will get you to your destination more quickly ninety-eight percent of the time, but two percent of the time you will be hit by a car. To make a rational decision between crossing on a red light or not, you have to take into account how good it is for you to get to your destination earlier and how bad it is for you to get hit by a car. A theory of decisions under uncertainty, based on the notion of utility, was developed by John von Neumann and Oskar Morgenstern, the founders of "game theory." In this theory, we assign a numerical quantity called a "utility" to each possible outcome. Rational decisions are based on combining the probabilities for various outcomes and the utility gained from each of these outcomes. The theory of von Neumann and Morgenstern has been the subject of intense debate in recent decades.

Thomas Bayes

Perhaps the major difficulty with the Bayesian point of view, whether relating to decisions under uncertainty or to the Bayesian philosophy of science, is that quite often, no one has a clue as to how to assign probabilities in cases of uncertainty.

The addition of probability thinking remarkably extends our understanding of reality. At the same time, we face the impossibility of understanding various phenomena, perhaps those which particularly spark our curiosity, despite having the language and tools of probability at our disposal. Introducing the language of probability allows us to ask many new questions that we cannot answer even using the tools of probability.

> Lena, Gina's beloved great-uncle, once told her: "The only thing I learned between the age of thirty and the age of fifty was how to deal with uncertainty." And then he sighed and added: "And the only thing I ever wished to learn after the age of fifty was how to deal with certainty."

5 Pre Review and Riskless Risks

"My own heart goes to Chemistry,"

Gina, Sept. 17, 2006

Gina Says:
September 17th, 2006 at 11:05 pm

And here is my review of last week.

"NOT EVEN WRONG", by Peter Woit; A review

(Replies and comments are very much welcomed, and this also means you, ReNorm)

Peter Woit is wrong claiming that "string theory" is "not even wrong." It is questionable if the distinction between "right", "wrong" and "not even wrong" coined by Wolfgang Pauli should be taken as a serious way of classifying scientific theories. It is a nice gimmick, though, and a great name for the book. In any case, the insights and truths offered by string theory, one of the most daring intellectual endeavors of our time, may well be wrong. They may also very well prevail as an important and unique part of physics. We cannot tell which way string theory will go.

"Landscape", the possibility of a huge number of theories that we may never be able to choose between, may be an artifact of string theory itself, or just of string theory in its present form. But it can also be an "impossibility result" which reveals a genuine problem with our ability to describe physical reality at some scales, whether we like it or not.

Whether string theory will prevail or not, it is already a well established fact that it has contributed important insights and technical infrastructure to mathematics and to physics

Even if "string theory" is the "theory of everything", string theory is not "everything". In mathematics we see many examples of such a distinction. Mathematical logic is a "mathematical theory of everything" that was developed in order to understand the foundations of mathematics. But while mathematical logic formally includes all other mathematics, in reality it is a beautiful field which is but one of many fields of mathematics and, as a matter of fact, a rather separate field. It took many decades before important links between mathematical logic and other mathematical disciplines were found.

While a theory which studies the most fundamental and general rules is appealing, in my mind, relevance is of key importance, which is why, of all sciences, my own heart goes to Chemistry.

Modern science and academic life do not give sufficient incentives for true scholarship. But Peter Woit is a scholar and large parts of the book exhibit both genuine scholarship as well as Woit's gifted ability to present and discuss in non-technical terms complicated mathematics and physics. Another advantage of the book is that Woit does not offer alternatives of his own to string theory. Woit does present a few nice ideas and observations that deserve to be pursued.

When it comes to string theory, Woit has concerns (some shared by string theorists), complaints (a few justified), suspicions, and unrealistic expectations (like everybody else); but Woit does not have a case.

Recommendation: For a layman wanting to read about string theory I would recommend Brian Greene's "The Elegant Universe" over Woit's new book. An intelligent reader should use grains of salt regarding any new scientific theory and any popular book describing it.

As I explained elsewhere, some of Woit's choices of content and rhetoric are disappointing.

I also do not like Woit's analysis of sociology, politics and funding of science. For a scientist, trying to explore something completely new (e.g., a replacement for string theory,) is a very very risky business. Woit aims at a system which allows scientists to take riskless risks. As there is no such thing as a riskless risk, Woit's ideas on this front may deserve the title "not even wrong".

Gina

(And thank you Ebgert for the analogy with mathematical logic.)

Riskless risk

The first reaction came from another Gina.

Gina Says:
September 17th, 2006 at 11:31 pm

Gina — do you have a background or degree in physics?

"Who is this Gina?" thought Gina and replied that she has some academic background but not a degree in physics. "The graduate" criticized Gina's review and said that it had to be more specific and support the points it made to a greater extent. Peter Woit himself was also negative and said: "I'm happy to debate people with a serious background in string theory who want to discuss the arguments in my book, but you're just wasting my time."

The "Renormalized" (or "born again Norm" as Gina used to think of him) asserted that the last paragraph "shows a total lack of understanding or insight into what Peter has said from the beginning", and continued: "Peter is not asking for a riskless risk, I feel dumb even copying 'riskless risk', he is asking for strings to be tied to reality in even the smallest way." Gina liked her "riskless risk" notion, and thought that in many respects, people in western society look for all sorts of riskless risks.

The long thread of comments was not over, and, as we will see in the next chapter, Gina had a chance to look back and have second thoughts about one of her claims.

Who is Gina?

As seen by Anat Amirav

Perhaps it's time to say a few words about our fearless Master of Ceremonies in cyberspace — Gina.

35 years of age, Gina is of Greek and Polish descent.

Born in the quaint island of Crete, she currently resides in the USA, in quiet and somewhat uneventful Wichita, Kansas. Gina has a B.Sc in Mathematics (from the University of Athens, with Honors), and a Master's Degree in Psychology (from the University of Florence, with Honors).

Currently in-between jobs (her last job was working with underprivileged children), she has a lot of free time on her hands, which gives her ample opportunities to roam the blogosphere.

Forever the proud Grecian, Gina is the happy owner of Papa, her beloved pet tomcat, named after "that dear man", Christos Dimitriou Papakyriakopoulos, whom she has been an ardent fan of ever since she first learned about him in high school. (See Chapter 7.)

Leaving behind Greece and Europe was not an easy decision to make. Still, Gina was certain that moving to the US would be a life-changing experience, which, if nothing else, would certainly do wonders for her less than perfect English.

Possessing an insatiable thirst for knowledge, Gina is a voracious reader and reads everything from scientific textbooks to People Magazine. She also enjoys writing, and keeps a daily journal. Since moving to the US, she has attended various journalism courses, eager to pursue her lifelong dream of becoming a professional writer. Until this dream is realized, however, she finds pleasure in expressing her musings on the blogosphere.

Gina is fascinated by both people and phenomena — the more complicated and controversial the better. An avid participant in the blogosphere, she has a genuine desire to understand the topics at hand, but, at the same time, being a fan of psychology, she would also like to investigate the character of the people who write blogs. She is even contemplating starting a blog of her own, but there's simply so much she'd like to say, she fears she wouldn't know where to begin, let alone end. Writing a blog, however, would certainly help her improve her English, which Gina knows is still not quite up to standard.

She is very inquisitive — a bit too inquisitive, according to some people. At times, her obvious tendency not to let go of a question or issue until she is absolutely satisfied can be quite exasperating. Gina is not oblivious to this annoying feature; it was actually one of the main reasons she decided not to pursue a career as a psychologist.

Gina, in fact, likes to think of herself as a modern day Lois Lane, who has yet to find her superman…any candidates???

6 Scientists are Paid to be Gullible

"My guess is that should individual scientists ever become 'objective and rational' in the sense of 'impartial and detached', then we should indeed find the revolutionary progress of science barred by an impenetrable obstacle"

> Carl Popper, "The rationality of scientific revolutions"

Gina Says:
September 20th, 2006 at 6:34 pm

One frustrating thing about this science business is the necessity to keep double checking not only those matters of disagreement, but also matters of complete agreement. There was a single item in the discussion I certainly agreed with Peter about, the need for an honest and self-critical approach, and even this item, on closer examination, is not that simple.

We do not pay scientists to be just honest, we pay them to be gullible, as well. They have to be gullible in their beliefs regarding the foundation of the current theory they are working on, the prospects for its success, the relevance of its methods, and its overall importance. And mostly they have to be gullible about their own abilities. They have to be somewhat blind to the frustrating nature of their profession, and to the overwhelming probability that, somewhere down the road, they may realize that they have been missing out on their lives.

Yatima Says:
September 20th, 2006 at 9:17 pm

Gina, this pertains to your interests, I think.

Just now I'm burrowing through Lee Smolin's book (in the middle of the night…hmm double special relativity? tasty!) but I will stop for now and quote from a novel by Arkadi and Boris Strugatsky (not to be confused with the Bogdanoffs) called 'Far Rainbow' which, on the face of it, is about a physical experiment going horribly wrong.

Towards the end, Camille, the greatest physicist of Rainbow, is heard to say (he has worked too hard, not to mention died, a few times):

"The Great Logician. Logical methods demand absolute concentration. To do anything in science, day and night you have to think about one and the same thing, read about one and the same thing, talk about one and the same thing... And where can you go from your psychic prism? Away from the inborn capacity to love… You've got to love, read about love, you've got to have green hills, music, pictures, dissatisfaction, fear, envy… You try to limit yourself — and you lose an enormous part of your happiness. And you know very well you're losing it. So then to blot out that conscious-ness and put an end to the torture of ambivalence, you castrate yourself. You tear away from yourself the whole emotional half of your humanity and you leave yourself with only one reaction to the world around you — doubt. Then loneliness lies in wait for you."

Hopefully most physicists are not so bitter. But still, one gets what he means.

"Who is Yatima?" thought Gina. Reversing the letters, one gets "Amitay." Perhaps Yatima stands for Amitay?

Drachmas (Taxi Story No. 1)

"The fixed price to JFK is 28 dollars" said the taxi driver; "toll and tips not included, and I want the two dollars and seventy five cents for the toll upfront." I reached for my wallet, dug eleven quarters and handed them to him. He carefully checked the quarters and said: "If you're wondering why I want the toll money here, it is all because of the Drachmas." "The Drachmas?" I asked. "Yes" said the driver. "They want to take me to trial for putting drachmas instead of quarters in the toll machine." Apparently, using 100 Greek Drachma coins, which are almost of no value, instead of US quarters created quite a problem. "No matter how much I tell them that I put whatever the clients give me in the machine they still do not believe me, and want to bring me to trial. Therefore I now check the quarters the clients give me here in New York, in the light." "I see" I said. I felt sorry for him. He was getting into serious trouble because of greedy, heartless passengers.

We started talking. The taxi driver told me about his old country back in Asia and how he came to the United States to start a new life. He is working hard driving a taxi, and makes a good living. And he has some plans and dreams. I told him about my country. We talked about family far away and about friends from the old country here in the States. His Drachmas story and the injustice done to him instantaneously connected us.

By the time we reached the toll machines we felt like old friends. "Look what I am doing" my new friend told me. He took my eleven quarters and put them in his right pocket, then pulled eleven drachma coins from his left pocket, showed them to me, and threw them to the basket. The gates opened and shortly afterwards we arrived at JFK.

7 The Poincaré Conjecture and Christos Dimitriou Papakyriakopoulos

"The perfidious lemma of Dehn

Was every topologist's bane

'Til Christos Papa-

kyriakopou-

los proved it without any strain."

John Milnor, around 1956.

Christos Dimitriou Papakyriakopoulos (based on a picture by Jay Goldman)

This chapter is about something quite different. Jumping from one topic to another, which is so characteristic of blog discussions, suited Gina quite well.

In 2002 Gregory Perelman, a Russian mathematician, presented a proof of the Poincaré conjecture concerning "three dimensional manifolds." Poincaré's conjecture was one of the most important problems in mathematics. Perelman's proof was an amazing scientific victory. His proof was rather concise, and filling up the details was itself an extremely important and difficult mathematical task, which several groups of mathematicians took upon themselves to carry out.

Woit's Sept 19 post was not about the Poincaré problem itself, but rather about the controversy surrounding Perelman's proof of the Poincaré conjecture, and certain remarks concerning it made by the famous mathematician Shing-Tung Yau.

This was the only time Gina really wanted to divert the discussion to a different direction, as she felt that the controversy regarding the solution of the conjecture was quite artificial, while the story of the Poincaré conjecture itself was so wonderful.

Is the story of the Poincaré conjecture related in any way to string theory? Not directly, but there are various indirect connections. The notions that Poincaré developed and, in particular, the notion of a manifold, became central in mathematics and in physics; the techniques used in the solution of the Poincaré Conjecture are closely related to physics. There is another connection related to the work of the mathematician S. T. Yau. String theory itself relies on the idea that in addition to the four dimensions we are familiar with (one dimension representing time), there are six additional dimensions related to a geometric object called a "Calabi-Yau" manifold.

Gina Says:
September 21st, 2006 at 7:08 pm

Wouldn't it be a better use of both time and space to talk about these exciting three dimensional manifolds themselves and how they are now understood, and what perhaps remains to be understood, rather than about lawyers and credits and newspaper articles and prizes and slanders and power struggles etc??

geometer Says:
September 21st, 2006 at 8:23 pm

Gina said: "Wouldn't it be a better use of both time and space to talk about these exciting three dimensional manifolds themselves..."

Well, the proof of Poincaré's Conjecture means precisely that there are no exciting (simply-connected closed) 3-manifolds: all of them are copies of the 3-sphere. It'd be much more exciting if the Poincaré Conjecture were false; unfortunately this is not the case.

Gina Says:
September 21st, 2006 at 8:39 pm

Actually, I have a specific question that maybe you guys can help me with. I vaguely remember the wonderful story of this humble mathematician whose nickname was "Papa" who worked on some things related to manifolds in dimension three, and who after years of effort managed to prove something really big. Thanks to Google and Wikipedia I found his full name — Christos Dimitriou Papakyriakopoulos, and apparently he proved the "Dehn's lemma". I am curious if the new proof of Poincaré's conjecture still relies on "Papa's" work.

geometer Says:
September 21st, 2006 at 9:18 pm

Gina,

I am not a 3d-topologist, but as far as I know Perelman's proof does not imply Dehn's lemma. I suspect that Dehn's lemma is used at the very last stage of Perelman's proof, when he obtains a collapsed 3-manifold and concludes that this must be a graph manifold.

Gina Says:
September 21st, 2006 at 10:35 pm

Dear Geometer,

Many thanks for the interesting information. I do not know what a graph manifold is precisely (never mind that), but I am very happy to hear that the proof of the Poincaré conjecture still relies on the work of that dear man "Papa", Christos Dimitriou Papakyriakopoulos. From what I heard he was a very special person.

You said, "It'd be much more exciting if the Poincaré Conjecture were false; unfortunately this is not the case."

I beg to disagree with you on this point. The way I see it, it is exciting that the Poincaré conjecture was proven true and it would have been exciting had it been proven false and, perhaps, the most exciting thing is that we could not have known in advance: not what the answer would be and not even if people would be able to crack this problem at all.

By the way, before going to dimension four, is everything known about manifolds in dimension two?

geometer Says:
September 21st, 2006 at 11:58 pm

Gina asked: "is everything known about manifolds in dimension two?"

Their classification is classical (pretzels with many holes and all that), but there are still some mysteries about surfaces, e.g. studying the mapping class group (i.e. the group of self-homotopy equivalences of a surface) is a very active area of research involving several branches of mathematics on which there is extensive literature.

Three-Dimensional Manifolds and the Poincaré Conjecture

Here is a drawing of three "two-dimensional manifolds without boundary"

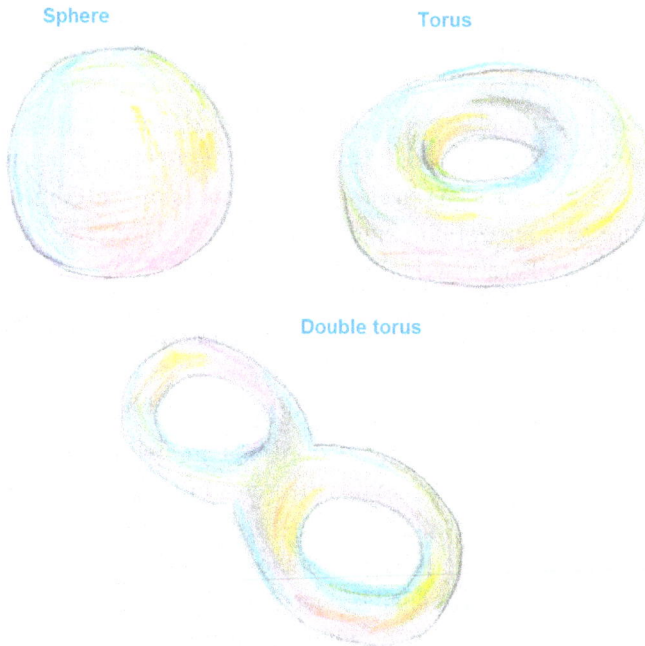

Sphere

Torus

Double torus

Every point on these geometric objects gives the "feel" of existing in a two-dimensional plane — just as every one of us gets the feeling that the Earth is flat. The sphere has another interesting property: every circle that you draw on it can gradually be shrunk to a point. However, this is not the case for the torus: a circle drawn around a "hole" cannot be shrunk to a point.

Three-dimensional manifolds are spaces with the property that around every point it "feels" like our three-dimensional space. In 1904 Henri Poincaré formulated a conjecture that suggested a characterization of the three-dimensional sphere as the only three-dimensional manifold "without boundary" in which every circle can be gradually shrunk to a point. (In 1900 Poincaré had made a stronger statement that he himself came to refute by constructing a space based on a platonic solid dodecahedron. This space is now called the Poincaré dodecahedral sphere.) People even formulated analogs to the Poincaré conjecture in dimensions higher than three. (The idea of dimensions greater than three is confusing. Bear with us and we shall return to these spaces in Chapter 12.) Quite surprisingly, the high-dimensional problems are easier to solve. Steve Smale proved the conjecture above dimension 4 and Michael Freedman proved it at dimension 4. The original Poincaré conjecture remained unsolved.

In 1982 William Thurston offered a much more general conjecture, called "the geometrization conjecture", that proposed a very daring geometric picture of all three dimensional manifolds. In the 1980s Richard Hamilton proposed a program for proving the geometrization conjecture and, in particular, the Poincaré conjecture. It was based on tools and ideas from analysis. Proofs of the Poincaré conjecture and of the entire geometrization conjecture, based on Hamilton's program, were described by Gregory Perelman in 2003.

The results of Christos Dimitriou Papakyriakopoulos still constitute a crucial part of the proof of the Poincaré conjecture. In 1956 "Papa" proved a result known as "Dehn's lemma", whose original 1910 proof by Dehn had a gap. (A few words of explanation may be of help. Mathematical results are called "theorems." Smaller results needed in the course of proving theorems are called "lemmas." In some cases proofs contain mistakes and often the mistakes are in those places where the authors feel that something is easy or obvious. In this case, proving the lemma that Dehn thought he had proved had been a major challenge for decades.) Besides Dehn's lemma, "Papa" proved two other fundamental related results called the loop theorem and the sphere theorem. For the rest of his life, "Papa" tried to settle Poincaré's conjecture, but did not succeed.

The next two chapters will describe Gina's efforts to understand some specific ingredients of string theory. What is the reason for the appearance of high dimensions? And why is it that so much effort is put into the study of two-dimensional models? What is the meaning of the infinite sums that are so useful in physics and in string theory?

8 Useful Divergence

Gina felt ready to ask a few questions about string theory, the magical high dimensions of our universe and the mysterious "infinite sums" — they were all so fascinating and cryptic.

Gina Says:
September 25th, 2006 at 10:24 pm

Here are a couple of questions I had while reading NEW ("Not Even Wrong").

1. String theory in its first simplest version implies that the universe has 26 dimensions. Is it possible to explain why in a few sentences, or a couple of paragraphs?

2. In the book there is a distinction (for very successful theories from physics) between "convergent sums","useful divergent sums", and "useless divergent sums". Is there any formal distinction between the last two types? Does it make sense in math?

3. The era before QCD and the standard model is described as being at least as confused and chaotic as the situation in string theory today, if not more so. Yet people who promoted these unsuccessful but very dominant theories were not asked to admit failure; moreover, they did not fail: conceptual and technical ideas from these unsuccessful theories turned out to be useful later, and students of these scientists had a crucial role in developing more successful and completely different theories. This looks like a good model to proceed with, doesn't it?

Gina was very thankful for Peter's answer.

Peter Woit Says:
September 25th, 2006 at 11:12 pm

Gina,

1. Away from d = 26, quantization introduces an "anomaly" in the symmetry of conformal transformations of the string worldsheet. Basically this means that the quantum string theory is more complicated and the metric on the world sheet becomes a dynamical variable you have to deal with. People study these "non-critical" string theories also. There are various different calculations that give you the 26, I don't know of any simple physical explanation for it.

2. Some divergent sums are "asymptotic" approximations to some function, which means that, at a fixed order, the truncated series is a better and better approximation to the function as the expansion parameter gets small (even though, at a fixed small parameter, as you go to a higher order, the sums sooner or later diverge). The perturbation series for QED is supposed to be such an asymptotic expansion. This kind of divergent sums can be quite useful, giving very good approximations.

3. Actually, the "bootstrap program", which was a dominant research program before QCD did fail as a theory of the strong interactions, and was pretty much killed off by QCD. Many of the people who worked on it certainly admitted that it failed; other die-hards never did admit this, but were no longer taken seriously by most theorists. Certainly some things learned from this were ultimately useful, but the program failed to do what it was intended to, and sensible people admitted this.

Gina learned about the "bootstrap program" from Woit's book. It was a dominant research program in high energy physics in the 1960's and overall it failed to achieve its ambitious goals. But she could not share Peter's interpretation of this story. This line of research did contribute some useful insights as Peter said, but much more than that, the students of those researchers who promoted the ideas that had failed, were among those who developed the theories that ultimately prevailed.

Convergent and Divergent Infinite Sums

The fact that the sum of infinitely many numbers can still have meaning is one of the most important insights of mathematics. A famous example is

$$1 + 1/2 + 1/4 + 1/8 + 1/16 + 1/32 + 1/64 + \cdots = 2.$$

The three dots mean that we continue forever. A sum like this one is referred to as an *infinite sum*, and when we can attribute a number to an infinite sum, the sum is called "convergent."

Convergent infinite sums are related to the old question of Achilles and the Tortoise. In this paradox Achilles is racing a turtle that has been given a 100-meter head start by Achilles. It takes Achilles some time (10 seconds, say) to reach the turtle's starting point, but during this time the turtle has advanced further, say by another 50 meters. (This is a quick turtle especially trained to demonstrate the infinite sum above.) It takes Achilles 5 seconds to reach the turtle's new position, but by then the turtle has advanced further still. Thus, whenever Achilles reaches a point where the turtle was, the turtle has advanced further and Achilles can never reach it. This is the paradox. The insight that an infinite sum can express a finite number is rather important in understanding this paradox. (Some philosophers disagree, claiming that the infinite sum description merely gives us a way to state the paradox, but not to solve it!)

Sometimes, convergent infinite sums are useful for computations. There are famous infinite sums that converge to the number π, thereby providing a very good approximation for π by computing just the first terms. In other cases, obtaining good approximations may require too many terms.

There are also infinite sums that are not convergent and they are called "divergent."

The sum

$$1 + 1 + 1 + 1 + 1 + \cdots$$

is divergent. A more subtle example is the sum

$$1 + 1/2 + 1/3 + 1/4 + \cdots$$

This sum is also divergent. How can we see that this is the case? One way is to note that the first term is larger than 1/2. And the sum of the next two terms (1/2 + 1/3) is larger than 1/2. And the sum of the next four terms (1/4 + 1/5 + 1/6 + 1/7) is larger than 1/2. And so on.

One of the surprising facts of modern physics is that certain divergent sums are still useful for computations and for theory. In "quantum electrodynamics (QED)," which is often referred to as the most successful physics theory ever (in terms of its precise predictions), the sums are described by gadgets called Feynman diagrams. Computational methods that are yet to be understood by mathematicians are common in most areas of physics.

Giving Meaning to Divergent Sums (A bit harder)

The sum

$$1 + 1 + 1 + 1 + 1 + \cdots$$

is divergent. Can we still give this sum a meaning? And what about

$$1 + 1/2 + 1/3 + 1/4 + \cdots?$$

It turns out that there is a context where we can associate numbers with these divergent infinite sums. The context is related to one of the most famous problems in mathematics, the Riemann hypothesis. The Riemann zeta function is a function which is described at any number using an infinite sum. It turns out that the function can be defined using a different approach even for values where the sum diverges! The bottom line is rather counterintuitive:

$$1 + 1 + 1 + \cdots = -1/2$$

and

$$1 + 1/2 + 1/3 + \cdots = -1/12$$

Unlike values of convergent infinite sums, we have to be quite careful in using values of divergent infinite sums. Nevertheless, they are important both in mathematics and in physics.

Another surprising fact is that in physics computing the first few terms of a "useful divergent sum" can, as the name implies, yield extremely good results, even though the entire sum is divergent. There is no clear mathematical understanding of this remarkable phenomenon.

9 Why Just Two-Dimensional Models?

Gina, you and quite a few others seem confused about the meaning of higher dimensions.

Thomas Love, September 28th, 2006 at 2:16 pm

From very high dimensions Gina moved to think about just two dimensions. Peter Woit mentioned the great successes of physics in two dimensions; "we should be trying to understand four dimensions" he wrote. Gina asked around about it.

Gina Says:
September 27th, 2006 at 8:18 pm

WHY THE EMPHASIS ON TWO DIMENSIONAL MODELS?

Peter wrote (in the book): "It is the precise expression of the mathematical relationship of representation theory and QFT that has been worked out in recent decades in two dimensions, exactly the thing that I would argue we should be trying to understand in the physical case of four dimensions."

Indeed one very nice point raised in Peter's book is the fact that many of the successes of physics and mathematics related to string theory and earlier physics are coming from two dimensional models. I asked around among my friends:

"Why can't you do anything as impressive for D>2, after all nobody, not even string theorists claim that our universe has two dimensions?"

It seems that for D>2 scientists are simply stuck and things look very gloomy. "Aren't there any ideas around," I asked. Well, there are a few. One guy told me with a spark in his eyes about an idea to move directly from $D = 2$ to $D = 4$ and to base models on "homological" notions which will extend important "duality" properties for planar models. He talked about things like "Poincaré duality" and "signature", and was quite excited but then admitted that these were all just ideas. (I guess this is the same old Poincaré.) Another guy had much hope for extensions to higher dimensions of "conformal analysis," which is prominent for two dimensional models using "representation theory." But this is also in a very premature state. A third guy praised the "Heisenberg Lie group" as a place to "be in high dimensions and to feel in two dimensions."

There are ideas, but overall there is also some feeling that studying higher dimensional models is a waste of time. Some of these guys actually spent a lot of time on it and got nowhere.

I tried to be tough on them and I asked if sticking to the cozy $D = 2$ and looking for the coin under the lamp is all about the summer salary.

"No, Gina" they said "this is not the reason". They said they simply do not know what to do. They need something to start with, perhaps a tip of a string to hold to, but often it turns out they just try to pull themselves up holding their own shoe laces. They have nothing, they said. They are quite savvy in failures but doing D>2 leads to "not even a failure." They did sound convincing but you never know with these wise guys.

Thomas Love Says:
September 28th, 2006 at 2:16 pm

Gina Says:

WHY THE EMPHASIS ON TWO DIMENSIONAL MODELS?

As a PhD mathematician, I know the answer: because the math is easier there.

Gina, you and quite a few others seem confused about the meaning of higher dimensions. Think of a dimension as an entry in an inventory form; how many numbers are required to describe the situation? (The standard reporter questions). Clearly, we need to know where and when (x,y,z,y), but we also need to know the field strengths, electric, magnetic, gravity, etc. Each of those requires another entry in the form, i.e. a dimension.

Gina Says:
September 28th, 2006 at 3:25 pm

"Gina, you and quite a few others seem confused about the meaning of higher dimensions."

You bet! I am very confused!!!

HEX and Percolation

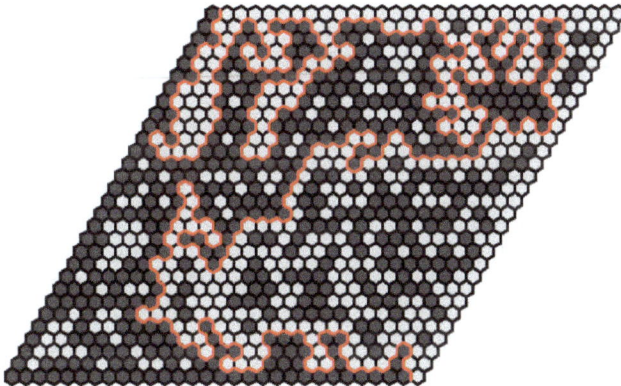

Here is a 30 by 30 board of the game HEX. In the original game one player, the BLACK player, fills an empty hexagon with black and the other player, the WHITE player, fills an empty hexagon with white. The WHITE player wins if there is a continuous all-white path from left to right. The BLACK player wins if there is a continuous all-black path from top to bottom. If we fill the hexagons at random with the colors black and white we obtain the two-dimensional statistical physics model of *percolation*. In the above picture all hexagons outside the boundary are colored randomly. The lowest black-white "border" (marked in the picture) is an important example of a stochastic curve in the plane.

Returning to the original game of HEX, John Nash proved a theorem that the first player has a winning strategy. His proof is not constructive and so nobody knows what this strategy might be.

10 Can Philosophy of Science Help Doing Science?

Enough with technical mathematics thought Gina. As the title of Woit's October 4 post was "Controversy, Controversy..." it was a good time for Gina to look at the philosophical arguments against string theory. These philosophical claims were at the core of the arguments against string theory, and they were not difficult to understand.

Gina Says:
October 5th, 2006 at 10:07 pm

Some remarks on the controversy:

DO WE REALLY HAVE A CONTROVERSY (YET)?

A striking fact about the debate concerning string theory is that there is almost a complete agreement on factual matters, between what string theorists say and what people who attack string theory say. The interpretation of the facts is sharply different, but many of the issues concerning the interpretation are not specific to string theory, and are of a very general nature.

CAN PHILOSOPHY OF SCIENCE HELP DOING SCIENCE?

This is a fascinating aspect of the discussion here and in Peter's and Lee's books. Philosophers will probably be the most skeptical about such "practical" applications of philosophy, for example, about Popper's point of view. It is hard to consider the various theories in philosophy of science as normative, and it is hard to consider them as descriptive. (They also are in conflict, of course.) These theories can be regarded as a way towards understanding and discussing what science is in a scientific way.

IS STRING THEORY FALSIFIABLE?

Well, I am not sure it is clear what string theory IS. But from the rough description of what it is, it seems very clear that string theory is falsifiable. For example, as Peter explained in his book, the 26-dimensional model without super-symmetry was rejected because it has consequences to physics that are regarded as unreasonable. This, in principle, can happen to the supersymmetric string theories.

DESTROY STRING THEORY AND SAVE PARTICLE PHYSICS?

Peter Woit's noble reason for his attack on string theory is the desire to save particle physics. Well, in general, as in this particular case, one has to be very skeptical about claims of the form "Destroy X to save Y". (But appropriate efforts to falsify or criticize a theory are as noble as efforts to prove or support it.)

Some Philosophy of Science

The Bayesian approach to the philosophy of science was developed in the first half of the twentieth century. Karl Popper and Thomas Kuhn are twentieth-century philosophers of science who later proposed alternative approaches.

It will be convenient to start with the Bayesian approach since we already talked about probability and Thomas Bayes in Chapter 4. The Bayesian approach (mainly associated with Ramsey and Savage) can be regarded as a verification-based philosophy of science; it is based on different scientists gradually updating, according to new empirical evidence, their (different) prior (subjective) probabilities of scientific explanations and theories, until the cumulative evidence is strong enough to reach a common conclusion.

One difficulty with the Bayesian approach is that in cases of disagreement, there are also disagreements on the interpretation of the evidence.

The Bayesian view does not offer a way to test a scientific theory but rather to update our beliefs in the theory given new evidence. In practice, scientific theories primarily explain existing observations. For example, the main motivation of Newtonian mechanics and the main support for its validity was the explanation of Kepler's laws. Kepler's laws concerning the elliptic orbits of planets around the sun were discovered seventy years before they were explained by Newtonian mechanics.

Karl Popper   **Thomas Kuhn**

Popper is famous for basing philosophy of science on the notion of falsification. According to Popper, the mark of a theory as scientific is falsifiability: the possibility to empirically refute the theory — in principle. This is in contrast with other approaches that can be viewed as basing philosophy of science on confirmation or verification. Famously, two principal examples of non-scientific theories according to Popper are the Marxist theory of capital and Freudian psychoanalysis.

If the Bayesian approach, like approaches based on verification, suggests that the optimal way for a scientific theory to proceed is by making safe conjectures which may lead to small incremental progress, Popper's approach suggests making bold and risky conjectures. One concern about practical implication of the Popperian approach is the fact that bold conjectures and theories that pass the falsifiability test are of little value if they are absurd or simply false to begin with.

Critics assert that neither Popper's theory nor earlier approaches based on verification give a proper description of how science is practiced. Also, they have limited normative value regarding how science ought to be practiced. It is especially difficult to use the insights from philosophy of science for scientific theories under development.

Thomas Kuhn is famous for his notions of paradigm shifts and scientific revolutions. According to Kuhn, science is normally carried out inside a certain paradigm that is shared by a community of scientists, and it is furthermore characterized by "paradigm shifts," which occur when the current paradigm is no longer capable of explaining the new evidence. Kuhn referred to the process of switching from the common paradigm to a new one as a "scientific revolution." An important example of a scientific revolution analyzed by Kuhn is the shift from Newtonian mechanics to Einstein's theory of relativity.

It may be useful to put things described here in a larger context and give a few more details.

Verification approaches towards philosophy of science: Verification (or confirmation) based approaches to scientific theories were developed in the first half of the twentieth century and became quite dominant in the philosophy of science, remaining so until today. An important role in their development was played by the Vienna circle, a group of philosophers, who shared common basic attitudes towards philosophy, and who gathered in Vienna mainly between the two world wars. According to most of these verification-based approaches, scientific theories are gradually examined and updated in view of new empirical evidence.

Philosophy and logic: The Vienna circle approach towards philosophy of science is part of their larger approach to philosophy centered on the rejection of metaphysics (and religion, in particular). This approach, called "logical positivism," is related to a greater move associated with Bertrand Russell and others to base mathematics, science and philosophy on logic. Logical positivism has led not only to distinctions of "what is scientific" and "what is unscientific", but also to an attempted classification of "what is meaningful" and "what is meaningless." As a result, for many decades the central interest in philosophy shifted away from traditional issues like "justice", "ethics", and "beauty" which were labeled as unscientific and even meaningless. (At a later time, attempts to distinguish "meaningless" from "unscientific" were made, and eventually the main interest in philosophy moved back to where it was before "logical positivism.")

Critiques of early versions of confirmation approaches in the philosophy of science were made, at the beginning of the 20th century, by French philosopher Pierre Duhem and by Henri Poincaré (whom we briefly met in Chapter 7). Duhem's approach asserts that the main test of a scientific theory is its internal coherence and consistency. Both Duhem and Poincaré gave much weight to "intuition" and "insights." This dispute is related to a famous debate between Poincaré and Russell on the role of logic in mathematics.

Probability: Probability plays an important role in some of the verification-based approaches to the philosophy of science. Often they rely on a logical (objective) notion of probability rather than "subjective probability" which is central to the Bayesian approach. Foundational questions regarding probability theory again come into play. Logical probability (also referred to as "partial deduction") is based on the idea that probability can describe a logical relation between two statements. This idea goes back to Wittgenstein, Caynes, and perhaps even to Leibniz. Rudolf Carnap, a central member of the Vienna circle, had a programme which he believed could lead to a whole logical calculus of probability starting with an answer to the question: "What is the probability of a statement A given the validity of statement B?" and ending with an answer to "What is the probability that a theory X is correct?"

Popper deliberately diminished the role of probability in his approach to philosophy of science. A (familiar) critique of Popper was expressed by Oded Schramm (who read an early draft of this book). Schramm wrote: "I remember when I read Popper, I was very disappointed with his treatment of probability. It was totally unsatisfactory. (This was much before I became seriously interested in probability.) Probability statements are never falsifiable."

It may be useful to put things described here in a larger context and give a few more details, and then to put them in even a larger context and give even a few more details, and then describe them in yet a larger context, and larger and larger …

Science is not only about philosophy and mathematics; it is also about reviews and criticisms. How to accept rejection? How to deal with criticism? How to relate to anonymous reviews? We will come to these topics in the next chapter.

11   Referees and Ethics

"I don't really think it's unethical to speculate on who anonymous reviewers are — I've certainly heard a lot of reputable scientists do so."

CapitalistImperialistPig October 10th, 2006 at 12:47 am

CapitalistImperialistPig

The title of Woit's new post was "**corrections...**" and one important item was a negative review of Peter Woit's and Lee Smolin's books by the science writer K. C. Cole. In trying to understand Cole's motives, Peter returned to the rejection of his book by Cambridge University Press and wrote: "As I said, I don't know Cole, so I don't know why she decided to write this kind of dishonest hit-job. Perhaps it has something to do with her professional association with string theorist Clifford Johnson at USC.

I've long suspected that Clifford was the author of the referee report for Cambridge University Press. One is often wrong about such guesses, probably I'll never know..."

Shortly afterwards an update was added:

"**Update**: Amazing how quickly one finds out things one thinks one will never know. Over at Clifford Johnson's blog, Capitalist Imperialist Pig <u>asked</u> <u>him</u> if he was the referee who tried to stop Cambridge University Press from publishing my book. His answer: 'that's all just silly and irrelevant'. OK, now I know..."

Peter continued: "The funny thing about this is that Clifford has been bitterly complaining about the fact that the book is being marketed and publicized to a wide audience, but it appears that he was the one who stopped it from being published a couple of years ago in a form where it would have reached many fewer people. Priceless."

In the discussion following these discoveries Gina expressed the view that Peter chasing the identity of the referee was rather unethical. The issue was not only about ethics, but also about interpretation of all sort of partial evidence, and how easy it was to become more and more confident about a certain factual matter from accumulation of very weak evidence. Peter Woit was increasingly confident that Clifford Johnson was the referee of his book, as he was increasingly confident that string theory is nonsense. Did he really have solid evidence?

Gina Says:
October 9th, 2006 at 8:21 pm

Peter and Tim,

If you are talking about the referee report that Peter is attributing (with little evidence) to Johnson, I think, in this case, ethics is going the other way around. Trying to reveal the identity of an anonymous referee, certainly in public, is not so ethical.

anon Says:
October 9th, 2006 at 9:06 pm

Gina, this sentence is quite fuzzy.

"Trying to reveal the identity of an anonymous referee, certainly in public, is not so ethical."

Who is trying to reveal?

Peter didn't ask Clifford. It was CIP who asked him. Are you accusing CIP of being unethical, for asking Clifford, IN PUBLIC? Clifford didn't have to answer, but he did answer. Peter FIRST quoted Clifford exactly, and then reached his own conclusions about the meaning of the quote. I don't think that is unethical.

I think for a reporter to write an article, with so many errors, and so one-sided in favor of her friend Clifford — now that is unethical.

Gina Says:
October 9th, 2006 at 10:01 pm

Dear anon,

Hmmm, interesting comment. I certainly do not "accuse" anybody of any-thing. For a scientist, trying to speculate in public about the identity of a referee is somewhat unethical, or perhaps a better word is unconventional.

Actually, Clifford did **not** confirm that he was the referee at all and he reacted in the appropriate way when he was asked — whether he was the referee or not — not confirming and not denying. So for Peter to reach a conclusion that Clifford was the referee, and to continue the discussion based on this assumption is not very good logic and not very ethical.

This is just a small point. There is also the issue of ethics and conventions for weblog behavior. This looks like a truly fascinating subject but I do not have anything to contribute. What can be the ethical rules for a Capitalis-tImperialistPig? I suppose whatever he does he cannot really be kosher.

Another 'Anon' came on board, with a capital 'A', and he agreed with Gina.

Anon Says:
October 9th, 2006 at 10:41 pm

I agree with Gina. It is definitely unethical to publicly accuse (or state that one "suspects") someone of being the anonymous referee.

That puts the accusee in the untenable position of being unable to either confirm or deny the accusation, and it is highly corrosive of the anonymous refereeing process (on which we all depend).

Peter Woit Says:
October 9th, 2006 at 11:41 pm

I'm not the one who put Clifford on the spot about this, blame that on CIP.

I've suspected that Clifford was the referee for quite a while now (since one day after seeing his behavior in response to challenges to string theory, and looking up over my desk and seeing a copy of his book on my bookshelf. I realized it was published by Cambridge, and a light dawned…).

What an interesting description of the process of 'discovery', thought Gina.

… I finally decided to mention this publicly because I really had enough with the way he was going on about what a money-grubbing publicity hound I am, promoting my book to the general public who can not understand the subtler points of what I am saying. If it weren't for him, the book would have been published two years ago by Cambridge University Press, in a form aimed at, and marketed to, a small audience. He decided to stop that, and left me no choice but to find a trade publisher. He has no business at all complaining about how this book was published, and it was unethical of him to do so, knowing full well that he was responsible for this.

I'm pretty sure I know who the second string theorist referee was, but won't say anything publicly about that. Unless he gets a blog and starts complaining about how the book was published....

Anon Says:
October 9th, 2006 at 11:56 pm

"I'm not the one who put Clifford on the spot about this, blame that on CIP."

Why?

You are the one who publicly stated the accusation (excuse me, the "suspicion"). CIP merely brought it to Clifford's attention, which someone else would have done sooner or later, anyway.

And then you were the one who took Clifford's non-committal response (the only ethical one he could make) as a confirmation.

CapitalistImperialistPig Says:
October 10th, 2006 at 12:47 am

I don't really think it's unethical to speculate on who anonymous reviewers are — I've certainly heard a lot of reputable scientists do so. Is it unethical to ask? It's a bit hard for me to believe that it is, but if it is I apologize, abjectly but not profusely, to the gods of ethical physics, if such there be. Generally speaking, the baloney I serve on my blog *is* strictly Kosher.

In any case, Clifford had several options — to ignore my question, criticize my question for violating the sacred bond between a publisher and the authors it already publishes, or to say something noncommittal. He chose the last option, but in a rather strange fashion — one that I thought looked like a politician's non-denial non-denial.

Marty Tysanner Says:
October 10th, 2006 at 1:13 am

Peter,

Having occasionally "defended" you in the past, I would feel a little hypo-critical in keeping silent on this. From what I have seen, I agree with Gina and Anon that you should not have publicly aired your hunches this way about Clifford's possible role as The Referee. It tends to create a suspicion of him without objective evidence. I also agree that you should not have publicly concluded that Clifford's non-answer to CIP showed that he really was that person.

Peter Woit Says:
October 10th, 2006 at 10:49 am

Marty,

Perhaps you're right. It's certainly true that I was highly annoyed by Clif-ford's recent behavior on his blog and because of that my judgment at the time may not have been the best. But still, I think mentioning my sus-picions can be justified, even beyond the grounds I gave earlier that Clif-ford should not be criticizing how this book was published, unless he is willing to have his own role in that story examined.

On this front, Gina thought, Peter's argument looked increasingly unrea-sonable. (She even thought it is completely legitimate for Clifford to regard Peter's book as unsuitable **both** for a university press and for a commercial publisher.) In any case, she did not expect any breakthroughs pertaining to Clifford's role. It was time to wrap up the refereeing issue and move to another topic discussed in the post.

Fundamental impossibilities

Understanding our fundamental limitations is among the most important contributions of science and of mathematics. There are quite a few cases where things that seemed possible and had been pursued for centuries, in fact turned out to be fundamentally impossible. Ancient geometers thought that any two geometric lengths are commensurable, namely, measurable by the same common unit. However, for a right triangle with equal legs, the leg and the hypotenuse are incommensurable. In modern language (based on the Pythagorean theorem), this is the statement that asserts that the square root of two is not a rational number. This was a big surprise in 600 BCE in ancient Greece (the story is that this discovery, attributed to a Pythagorean named Hippasus, perplexed Pythagoras to such an extent that he let Hippasus drown). Two centuries later, Euclid devoted the tenth book of his work the *Elements* to irrational quantities. The irrationality of the square root of 2 is an important landmark in mathematics. Similarly, the starting point of modern algebra can be traced back to another impossibility result. Algebraists found formulas for solving equations of degrees two, three, and four. Abel and Galois proved that no such formula is possible for general equations of degree five and above. This theory also led to a solution of a problem that remained unsolved from the Ancient Greek era: finding a method of trisecting a general angle with a compass and a ruler. Galois' theory demonstrated that no such method exists.

There are other important impossibility results. We have already mentioned Gödel's impossibility result, asserting that it is not possible to prove (nor even to state) the consistency of mathematics from within mathematics. It is believed that there are inherent impossibilities for computers (and for any computational devices). This is the famous conjecture that P is different from NP. An impossibility theorem of Gibbard and Satterthwaite asserts that for an election with more than two candidates and at least two voters, there does not exist a voting method; that is, there is no method for choosing a winner, based on the voters' preferences, which is immune to manipulation. Impossibility insights outside mathematics, like the impossibility of building a perpetuum mobile, of turning iron into gold, and (most likely) of traveling in time, are related to profound scientific understanding. And, of course, the impossibility of traveling faster than the speed of light is one of the famous and mind-boggling insights of science.

12   It is All a Fantasy: Cole's Review

"Simply put, string theory does this by replacing point-like particles with tiny strings of some fundamental stuff vibrating in 10-dimensional space; their harmonies creating everything from quarks to galaxies."

K. C. Cole, LA times, October 2006.

Vibrating strings create everything

"This is what you have never gotten in your head Gina. It is all a fantasy!"

renormalized, October 11th, 2006 at 9:55 am.

Gina thought it was time to pay attention to K. C. Cole's review itself. The truth of the matter was that Gina liked the review and did not even consider it to be so negative towards Peter Woit and Lee Smolin. Cole's review was entitled "Strung Along" and Peter Woit regarded it as "basically a hit-job on me and Smolin."

Gina Says:
October 10th, 2006 at 5:03 pm

I agree that the issue of refereeing is not simple and sometimes loaded. In fact, almost every negative referee report is a little controversy. A completely open system may be an option; a "double blind" system is another option. But I do not see a simple solution except the universal advice: "take it easy".

I read Cole's review and it does not look to me like it's that negative. It is critical both to string theory and to the books by Lee and Peter, and this seems reasonable. There were 20-30 reviews on the book, most are positive, a few (like my own) are negative. I do not think it is right or wise, Peter, to regard negative reviews as "hit jobs" and to "go after" the people who write them. If you want people to listen to your criticism you should be ready to accept criticism.

Arun Says:
October 10th, 2006 at 6:16 pm

[Quoting from Cole's review as quoted by Woit:] "Woit, and Smolin in 'The Trouble With Physics,' write mostly about how string theory has ruined their careers."

Whether or not a book is worth reading is a matter of opinion, and one can have any opinion. But what is in the book or not in the book should not be lied about.

Gina Says:
October 10th, 2006 at 7:39 pm

Arun,

Indeed this quote struck me as unfair but I double checked, and it is not the full quote. Here it is:

"These issues are well worth addressing, which makes it all the more disappointing that Woit, and Smolin write mostly about how string theory has ruined their careers — <u>and physics as well</u>."

I think it is correct to characterize Peter's position as claiming that string theory ruined particle physics, and this is what Peter mostly writes about in the polemic part of the book. The part about string theory ruining Peter's career is an interpretation by Cole of what Peter tells about himself in the book. This is what Cole reads between the lines. It is a legitimate interpretation (for a journalist).

anon Says:
October 10th, 2006 at 8:47 pm

Gina, I've never, ever heard Peter say that string theory ruined his career. In fact, Peter sounds to me as happy with his career as a pig in mud. Cole's sentence is groundless, a calumny, a character assassination; what she means is that Peter criticizes string theory because he is bitter; that his criticism has almost nothing to do with any shortcomings of string theory.

Capitalist Imperialist Pig Says:
October 10th, 2006 at 11:54 pm

Gina, your defense of KC Cole's review does not conform to a reasonable interpretation of the facts. Neither does her review. It wasn't merely biased, it was deeply dishonest. Writers (and actors, painters, and others) are totally justified in criticizing their critics, especially those who don't evaluate them honestly.

Since you have strongly insinuated yourself into this debate, including questioning my ethics, let me ask the following: what is your expertise? Are you a string theorist? A physicist?

Oh well, thought Gina. It was legitimate for Cole to speculate that Woit and Smolin are also driven by personal frustration. But even if this is the case it makes little difference. Personal motivations are often much overplayed. The main issue is the quality of Woit's and Smolin's arguments.

Gina Says:
October 11th, 2006 at 8:42 am

Well guys, apart from the controversial sentence regarding Woit's and Smolin's careers, Cole's review is good.

Take this one sentence explanation of string theory:

"Simply put, string theory does this by replacing point-like particles with tiny strings of some fundamental stuff vibrating in 10-dimensional space — their harmonies creating everything from quarks to galaxies. The loops of string don't let anything get small enough to let quantum fidgeting rip space and time apart."

And the critique on string theory:

"String theory has its troubles, which the authors analyze in great and sometimes lucid detail: It appears to be untestable because the strings are too small to be seen, and recent research suggests that the theory may have an infinite number of solutions, so it can't make predictions. And string theory is so ill-defined that even ardent supporters admit they don't know what, exactly, it is."

And

"The authors are right to say that physicists can get cliquish; that some of them swagger; that they frequently fool themselves and that science has become too risk-averse."

Bert Schroer Says:
October 11th, 2006 at 8:48 am

"Simply put, string theory does this by replacing point-like particles with tiny strings of some fundamental stuff…"

Well, this mantra with which string theorists used to start their talk is totally metaphoric.

Renormalized Says:
October 11th, 2006 at 9:55 am

"Simply put, string theory does this by replacing point-like particles with tiny strings of some fundamental stuff vibrating in 10-dimensional space — their harmonies creating everything from quarks to galaxies. The loops of string don't let anything get small enough to let quantum fidgeting rip space and time apart."

This would be wonderful if it were true. The truth is we have never seen a string, never had an experiment which inferred there were strings, have never seen a dimension beyond the 3 dimensions for space and the one dimension for time. We have never seen this so called "fundamental stuff". We have never heard the so called "harmonies". We have never found a loop of string. This is what you have never gotten in your head, Gina. It is all a fantasy! This is what happens when you get one good idea about how the universe "might" work and then have many brilliant people working on it for many years, teaching brilliant students to work on the idea and then forgetting it was just an idea that "might" work.

Gina had her disagreements with "renormalized" in the past and was often puzzled by his language and attitude. But she felt that, right or wrong, string theory, as many other highlights of human activities, represents wonderful fantasies. Speaking of his own life, which was so full of events, Gina's beloved great uncle Lena often told her: "It is all a dream, Gina; it is all a fantasy."

It is all a fantasy, Gina!

Many Dimensions

Planar and spatial geometry were the flagship of ancient mathematics, while algebra was the central mathematical creation of the Middle Ages. René Descartes merged these two subjects via his analytic geometry. The idea was to present geometric objects like points and lines by means of variables and equations. A point on the plane can be represented by a pair of numbers (x, y), called coordinates, that uniquely describe its position. A point in space is similarly described by a triplet of numbers.

An important notion in geometry is that of dimension. A point has dimension 0, a line has dimension 1, a plane has dimension 2, and space is three-dimensional. The universe we live in is usually described as a four-dimensional object, where the fourth dimension refers to time.

In the nineteenth century, mathematicians began studying various new notions of "geometry." Geometries that violate Euclid's "parallel postulates" were created; the most important of these is perhaps hyperbolic geometry. Finite geometries, namely geometries that contain only a finite number of points and lines (unlike the familiar Euclidean plane that contains infinitely many points and lines), were also created. Mathematicians also began studying geometries of dimensions higher than three. The basic idea is very simple: if we can represent every point in the plane with two numbers, and in our familiar three-dimensional space we need three numbers, in a geometry with N dimensions we can represent every point using N numbers. Here N can be any natural number 1,2,3,4,5,… (Mathematicians even considered geometries of infinite dimensions and geometric objects whose "dimension" is not a natural number.)

High-dimensional spaces carry many surprises. Ancient Greeks discovered the five regular polyhedra: the simplex, cube, octahedron, dodecahedron and isocahedron. These five geometric objects are also called the platonic solids. To the list of 5 platonic solids in dimension 3 we can add 6 hyper-platonic solids in dimension 4. Something magical occurs in dimension twenty four. A certain construction of a configuration of points in 24 dimensions discovered by John Leech in 1964 had an immense impact on several areas of mathematics.

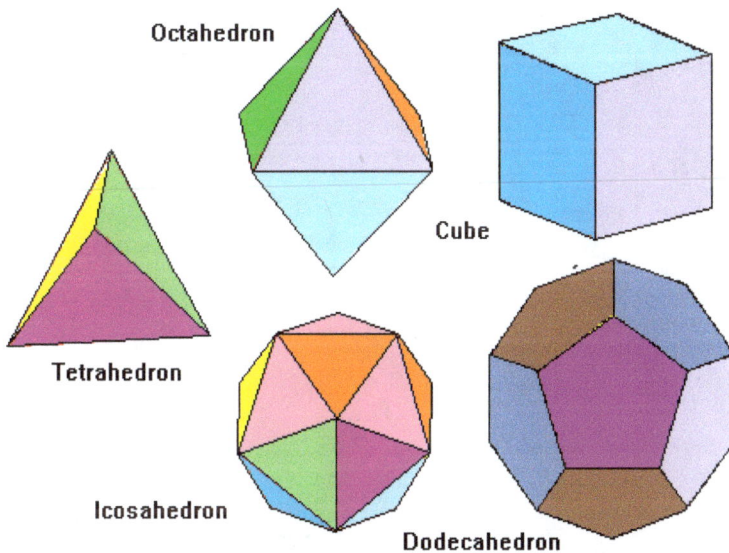

The five regular polytopes (platonic solids). 19th century mathematicians discovered six regular polytopes in dimension four.

Replacing the 3 + 1-dimensional space used to model our physical world (one dimension describing time) with a higher-dimensional space allows for new opportunities in modeling our universe and resolving difficulties in traditional theories. In the 1920s Theodor Kaluza and Oskar Klein proposed that a fourth spatial dimension will allow the theory of gravitation and electromagnetism to be unified. Moving to high dimensions has a central importance in string theory. The vibrating strings "live" in rather complicated ten-dimensional geometries. A further insight of string theory is that dimension itself need not be a firm characteristic of a physics theory: physics theories in different dimensions can lead to precisely the same rules.

13 Expelled

"Gina's comments are blocked on my blog because she was posting a large number of comments there, while most of the time clearly not understanding what she was writing about,"

P. Woit, "Backreaction" 5:59, December 27, 2006

Woit's next post titled "Rube Goldberg's Instruction Manual" again dealt with the "landscape" of the many different possible string theories. It was a high-level, quite technical, yet interesting discussion, and Gina offered a short summary. A few minutes later:

Peter Woit Says:
October 17th, 2006 at 9:06 am

Gina,

You're adding nothing to the discussion here, just repeating in garbled form what others who do understand the issues have written. In particular, Joe does not claim to have predictions common to all theories; he was discussing something that is a feature of one class of compactifications but not others. Please stop adding to the noise level here.

anonymous Says:
October 17th, 2006 at 9:58 am

I am just an intermediate grad student, but I agree with Gina.

As far as I can tell, the situation is: there are "old" string theorists who are stuck in 1985 and hope the whole thing [the landscape] will go away; more modern string theorists who realize this [the landscape] is a real issue with the whole theory (but the theory may well still be correct and one has to deal with the issue); and finally a bunch of people like Peter and Smolin who really dislike string theory, for reasons you can judge yourself (they don't seem very sound to me). Anyway I need to read my next review now (about supersymmetry), so I'll sign off.

Then this delightful graduate student was confronted by some savvy participants. For example, **Jean-Paul** asked: "Before you run away, I am just curious why you want to enter a research field which is in a state of crisis... What do you expect to learn?" Gina found Jean-Paul's comment unfair, but when she tried to post a little note of encouragement she discovered that her messages were blocked. She checked the matter with Peter and realized that she would basically not be able to post further remarks. The next post by Peter titled "**The String Wars**" was an appropriate place for her farewell.

Gina Says:
October 18th, 2006 at 6:50 pm

Goodbye!

This is a little off-topic, but I hope Peter will allow it, just this time. Over the last eight-ten weeks I made occasional comments on several issues discussed here and in Peter Woit's book. I did put a lot of thought into my comments, although, I could not always be on par with the blog speed. My comments were mainly on issues regarding philosophy of science, and the practices and ethics of debating science. Following my attempt to summarize the recent landscape discussion, Peter informed me that since my contributions were drawing lot of criticism from other participants, criticism that he shares, I will not be able to post here without prior monitoring. This basically means that I will not be able to comment here anymore.

It was an interesting experience, even if unsuccessful. Best wishes everybody.

Peter Woit Says:
October 18th, 2006 at 7:37 pm

About Gina's comment: As the volume of comments has increased here, it has become more and more difficult to keep the noise level down… especially difficult is the phenomenon of people who sometimes post something sensible and interesting, but all too often something that isn't.

ksh95 Says:
October 19th, 2006 at 10:59 am

Gina,

I have a Ph.D in physics, yet I would never be arrogant enough to go to n-category café [a name of a blog], start posting uninformed nonsense, jump in the middle of discussions I can't fully understand, and then sulk when my posts get deleted. **Instead**, I would read every day, try to learn as much as I could, and feel lucky that I was privy to such a high level discussion. Any post I would make would be to thank the blog owners for making such interesting discourse public, OR, to ask for simple lay-men explanations.

Hal Says:
October 19th, 2006 at 6:34 pm

To add something missing from ksh95's comment I would like to wish Gina well since she did, after all, give her best wishes to everyone.

ak had a different opinion.

ak Says:
October 20th, 2006 at 12:45 pm

…The more informal and liberal form of blogs makes it possible to share information flow between possibly poorly overlapping subjects and standpoints and people with very different knowledge backgrounds. This cannot be regarded as a pure 'noise phenomenon'.

The scientific blogosphere can serve the purpose of physics and math-ematics, classically highly impenetrable subjects for non-experts, to com-municate with the 'outer world'. Radically differing forms of reasoning expressed by philosophers, sociologists or even artists may be engaged in informal discussions with the mathematical or physical world. Such an interaction could be of benefit to physicists or mathematicians themselves.

To be more concrete: as it seemed to me, Gina, for instance, does have a scientific background, and while it apparently does not quite overlap with particle physics (chemistry?), it does not quite exclude her from any form of rational reasoning, either.

Peter was, however, determined:

Peter Woit Says:
October 20th, 2006 at 2:11 pm

ak, sorry to be a bit obnoxious about this, but one of the whole points of
the blog technology is to allow people to try and create an information
source and discussion forum of whatever kind they want. One obvious
thing to point out to people who don't like this is that the technology
is free, you're welcome to create your own blog, and there you can do
exactly what you want.

Noise

What is the correct picture of our world? Are noise and errors part of the essence
of matters, making the beautiful perfect patterns we see around us, as well as the
notions of information and computation, just derived concepts in a noisy world?
Or do noise and errors just express our imperfect perception of otherwise perfect
laws of nature? Talking about an inherently noisy reality may well reflect a better
understanding across various scales and areas.

14 H. S. M. Coxeter, Polytopes and Mirror Symmetry

The next post on Woit's blog was about the mathematician H. S. M. Coxeter and his geometry, with little connections to string theory mentioned. D R Lunsford shared his memories of reading Coxeter's famous book on regular polytopes:

D R Lunsford Says:
October 19th, 2006 at 3:54 pm

"I remember reading 'Regular Polytopes' as a kid and discovering empirically a fact about the Platonic solids. I was eating a lot of Dannon Yogurt at the time — the container was environment-friendly wax paper which was nonetheless rather weak. To strengthen the top, a circular cardboard disk was inserted. I pried out a bunch of these identical disks and used them to make the five solids by inscribing regular polygons in them etc. In the end, each face of each solid could thus be inscribed in the same circle. When I set them on my desk, I noticed that they paired up in altitudes, the cube and octahedron having the same altitude, likewise the icosahedron and the dodecahedron, while the tetrahedron was paired with itself, being self-dual!"

Reading this, Gina, who was also fascinated with the dodecahedron and the icosahedron, was tempted to submit one last comment.

Gina wanted to say:

"(Just one more remark for the road…) It is hard to avoid connections with string theory. The dual pairs of polytopes that DR Lunsford talks about (but in high dimensions) are undoubtedly connected to string theory mirror symmetry. And the string vacuum project offers to study all sorts of polytopes, objects that Coxeter apparently loved. (Another reason for NSF to support the project!)

The comment was waiting for moderation for a long while, and then Gina, for the first time, e-mailed Peter and asked him about it.

> Gina,
>
> you are wasting far too much of my time that I can't spare. This comment was uninformed and I'm really tired of having to spend time here or on the blog discussing this. For one thing, the reason algebraic geometers have been so excited about mirror symmetry is that it's a far more subtle and unexpected phenomenon than just the kind of geometric duality of polytopes that you write about.
>
> Peter

Gina did not fully agree, but felt that Peter may have had a point and that her comment was too brief and cryptic. She submitted a more detailed comment which she thought was a nice contribution to Peter's blog, and she also wanted to impress Peter a little.

Gina wanted to say:

> I did not plan on making more remarks on this weblog, but then Coxeter came along, and, you see, not only have I always had great admiration for him, I even once met him and his wife. I would like to tell you a story that involves polytopes that Coxeter studied, and love, and string theory.

Gina continued to talk about polytopes, and about duality of polytopes which is connected to string theory mirror symmetry. She then spoke about other beautiful mathematical objects called "partitions" and about another nice and surprising chapter in the understanding of mirror symmetry discovered by the recent Fields medalist Andrei Okounkov, and several partners.

A partition is just a way to write a number as a sum of other numbers. Like 10 = 5+3+1+1. You can represent the partition by a picture (diagram) like this:

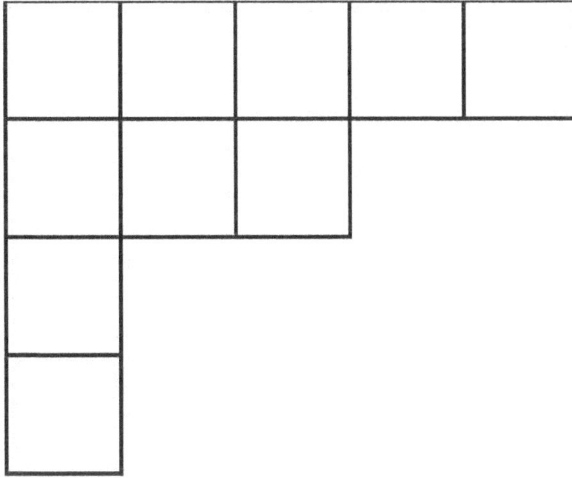

(This is a diagram representing 10 = 5+3+1+1.)

Partitions have attracted mathematicians for centuries. Among others, the famous Indian mathematician Ramanujan was well known for his identities regarding partitions. And now enters another idea, bearing the names of Ulam, Vershik, Kerov, Shepp and others who studied partitions as stochastic objects. In particular, it was discovered that "most" partitions, say of a number n, come in a "typical shape".

The emergent picture drawn by Okounkov and his coauthors goes very roughly like this: an "algebraic variety" (a manifold of some sort) that takes part in a certain string theory is related to a class of partitions, and when we consider the typical shape of a partition in the class this gives us another algebraic variety, and — lo and behold — the typical shape IS the mirror image of the original one. The mirror relations translate to asymptotic results on the number of partitions, somewhat in the spirit of the famous asymptotic formulas of the mathematicians Hardy and Ramanujan for p(n) — the total number of partitions for the number n.

Peter remained unimpressed, and the remark was not accepted for publication.

And **apropos** Larry Shepp, one of the heroes in the story above, Gina recalled the following amusing story. Shepp gave a talk at Columbia University and presented a new statistical model. Transparency 3 of his lectures ended with the claim:

"This model is as profound and as interesting as any model in statistical physics..."

This claim was regarded as an insult, and caused some discomfort in the audience that soon turned into a heated debate. The lecture was interrupted for 10 minutes and the situation nearly escalated into a fist fight. When things calmed down and the lecture resumed, Shepp displayed the next transparency, which began with the phrase:

".... well, perhaps not quite."

Larry Shepp

15   The Future of String Theory

"Utopia — 10%, Triumph and isolation — 40%, perpetuum mobile — 10% decay — 20% failure — 15% alternativa — 15%,"

Gil Kalai [trying to add up to 100%], Nov. 16, 2006.

As in any separation, Gina felt that she was certainly also to blame. She was at times, rather annoying. She was a little shocked to be expelled but did not have hard feelings. For one thing, she did learn a lot from Peter's book and blog, and looking at the world mainly with curious eyes was also a means to avoid hard feelings. Being a cyber entity she found thoughts about her *feelings* quite confusing. Overall, Gina liked Peter. When he did not discuss string theory he could be very interesting. Peter also had an endearing feature of fighting fiercely when he was not appreciated, and at the same time, becoming embarrassed and overly humble when he was. Gina thought it could have been better for Peter to receive compliments more willingly, and to accept criticism more lightly.

Indeed, Gina was rather critical of Peter Woit's skeptical approach towards string theory and referred to it once as "skeptical harassment". She wondered what the role of skepticism should be in science, and what should be a correct way to pursue a skeptical point of view. This is a fascinating question, she thought.

Gina had a few comments left in her drawer. One was about debating beauty. Is it at all possible? Another was about the "future of string theory." She was pleased to see that somewhat similar ideas were expressed by another participant.

Gil Kalai Says:
November 16th, 2006 at 6:37 pm

When brilliant and hilarious Scott Aaronson came to town last month, he was much more eager to tell us about the recent physics controversies, the new books about string theory, and the related blog excitements, than to discuss quantum computer's skepticism. Scott surely got us interested!

What might happen to physics in the next 50 years? Good question! And what does the future have in store for string theory? Below are six alternatives:

This (light) piece is inspired by the (deep and serious) classic paper by Russell Impaglliazzo on the five possible universes of computation and cryptography.

Apart from the illustrative details (which are meant to be amusing), I regard each of the six alternatives below as realistic. The second alternative can be regarded as the current default cautiously-optimistic mainstream approach of the scientific community which, perhaps, makes it the most plausible option.

Six Alternatives for String Theory's Future

1. UTOPIA

String theory continues to progress and converges in a few decades to become a solid part of our scientific understanding with plenty of empirical direct and indirect confirmations and many applications to all other areas of physics. Some of the landmarks after the "Maldacena conjectures" (1997) were the "Johnson Postulate" (2009), the "Motl Ansatz" (2014), the "Distler Paradigm" (2017) followed by the powerful "E-F-W Calculus" (2022). String theory becomes the "language of physics", perhaps even "the language of nature". Every graduate student in physics is able to make string theory computations, and this is what most physicists do. String theory represents a sound mathematical theory, in fact, mathematics is now considered as just "the special case of string theory for Plank constant 0". A few exciting problems remain.

Peter Woit's book "TRUE!" tops the NYT best sellers lists for 24 weeks. (In his book Woit advises caution when applying string theory to the area of finance.)

2. TRIUMPH and ISOLATION

String theory continues to progress and converges in a few decades to a solid part of our scientific understanding with convincing empirical direct and indirect confirmations, but with little applications and relevance to other areas of physics. Computations with string theory are extremely hard. (Computations based on the E-F-W calculus are computationally infeasible even on the newly built "quantum computers".) Mathematical foundations of string theory, much like those of earlier high-energy physics, remain shaky.

Peter Woit's book "NOT WRONG!" hits the market.

3. PERPETUUM MOBILE

String theory continues to progress but it does not converge. String theory thus remains a "useful divergent theory", whatever this means. More and more exciting connections to mathematics are found. More and more conceptual revolutions in the theory itself are taking place. (The latest is the "13th superstring revolution".) String theory leads to an entirely new way of looking at physics and, even more so, it is a scientific experience not seen before. As always, the best, most brilliant minds are attracted to this theory.

The 17th edition of Woit's "Not Even Wrong" appears.

4. DECAY

String theory continues to progress but the progress is slower; the attractiveness of the theory seems to diminish. String theory still looks promising, but while the success appears to be just around the corner, string theory is not sufficiently promising to attract the best people. Interest in physics is shifted to other directions.

5. GLORIOUS FAILURE

A brilliant string theorist from Vanderbilt University discovers a potential feature of supersymmetric string theory which contradicts basic physics insights. Massive computations in the "String Vacuum Project" confirm her discovery. After several years of extensive research (with beautiful new connections to mathematics found) it is now commonly accepted that string theory was falsified, and is no longer an option for a theory of everything. No alternative is in sight. 20 prominent string theorists declare string theory as part of "mathematical physics", rather than a viable physics theory, and within 72 hours, 18 of them get lucrative offers from top mathematics departments.

Woit's biography of Ed Witten "WRONG!!" is the basis for a successful Hollywood movie featuring Will Smith as Witten in the main role.

6. ALTERNATIVA

The alternative theory was discovered by sheer coincidence and, like string theory itself, is based on a technical rather than conceptual idea. The initial step was made by an elderly mathematician from Bristol University who, seeking a mathematical explanation for QED, suggested a small correction to the Feynman diagram expansion. Strangely, this has led to some consistent theory and made quantum gravity easier. The next step came when a researcher from the University of Tehran (provoked, in parts, by some rather general suggestions of P. Woit, and the mathematical notion of "noise sensitivity",) connected dark matter and dark energy with representations of unbounded weights and dimensions. Such representations are prominent in the new theory. (This new type of mass/energy is called "the mess".) The theory subsequently developed and was brought to completion by New-Jersey based physicists.

An extremely surprising feature of the new alternative theory is that the universe is 3+1 dimensional.

The translation by Lubos Motl of the new edition of Woit's "Not Even Wrong" to Czech has just appeared.

(My subjective probabilities for the future of string theory, Utopia —
10%, Triumph and isolation — 40%, perpetuum mobile — 10%, decay —
20%, failure — 15%, alternativa — 15%. Of course, some combination or
entirely different scenarios that I missed are also possible.)

One blogger immediately noticed that those subjective probabilities add up
to 110%. Oh well.

Reading this, Gina recalled the story about Ulam and the future of
mathematics:

Ulam and the "Future of Mathematics"

Ulam was scheduled to give a talk at the University of Chicago titled "The future
of mathematics." Stanislaw Ulam was a rather famous mathematician and a major
player in building the H-bomb, so a large audience gathered.

Ulam had trousers with two suspenders converging at a single front button.
At some point in the talk, when Ulam became excited, this single button loosened,

as did the suspenders and the trousers came down slowly, revealing colorful boxer underwear. Ulam did not notice it for a couple of minutes and then when he did, he raised the trousers and just supported them with his hand, which was fine, except that when he got over-excited and waved both hands, the trousers came down again.

Besides the colorful boxer under-pants, Ulam's lecture was a completely mundane mathematical talk: some personal memories of him and 'Johny', along with problems, lemmas, theorems, conjectures, little proofs, nothing unusual, and not a word about the future of mathematics.

When it was time to ask questions, a person from the audience asked Ulam about the future of mathematics. Ulam looked at him rather surprised, and replied very slowly.

"Young man," Ulam said, "you must understand: we are now in the present, and the future only comes *later*, so it is not possible to know what the future of mathematics will be, simply because it has not happened *yet*."

Then other people asked similar questions and reminded Ulam that the title of the talk was "The Future of Mathematics", but Ulam insisted that these questions were misguided since the future would only come at a later time.

16 Interlude: Silly String Wars in Jerusalem

"And, finally, of course, 'string wars' is mine. All mine!"

Aaron Joseph Sturmthal Bergman
[Claiming ownership of the term 'string war'] 11:37 AM,
December 24, 2006

"Silly string wars are a common way of celebrating the day in Israel as seen on Ben Yehuda Street in Jerusalem on the holiday in 2006"

Yom Ha'atzmaut — Wikipedia, the free encyclopedia

Since the mid eighties, silly string wars have become a popular way to celebrate Israel's Independence Day (Yom Haatzmaut). Initially rooted in the competition between two prominent Jerusalem high schools, the "Leyada" and the "Gymnasia", silly string wars are nowadays conducted all over Israel. Every year, on Independence Day, festive young men and women split into groups of string opponents and string proponents and engage in the fun.

Among the silly things opponents do, is silently approach a group of proponents and surprise them with a loud shout "**t' Hooft**!!".

Silly proponents, on the other hand, follow their leader, who in turn asks: "What are we doing?" "Add" "and what is the answer" "Ten" "and what do you get?" "Add-we-ten" "Add-we-ten", the ecstatic proponents cry cheerfully. They also take out small mirrors, place them next to bystanders' faces and shout: "mirror symmetry!"

You can hear head-to-head clashes between proponent and opponents who, referring to the finiteness of string theory, incessantly shout at each other: "Finite" "infinite" "finite" "infinite..." In recent years these roles have occasionally been reversed when proponents and opponents, referring to the finiteness of $N=8$ supergravity, shout endlessly: "infinite" "finite" "infinite" "finite" "infinite..."

When the festivities end, people go back to their homes. They are ready for yet another year of serious productive activities in the boring city of Jerusalem, where nothing ever happens.

Part II

The n-Category Café

Where to go? What to do now? Gina asked herself. All in all Gina was not satisfied with gallivanting and jumping to and fro between different blogs. She wanted a blog community to belong to. Back on Woit's blog somebody called ksh95 mentioned the n-category café. The n-category café seemed like the worst possible choice. Gina did not even know what n-categories were, and she could not follow most postings. Yet the people there were nice and Gina decided to go for it.

The n-category cafe is a group blog on "math, physics and philosophy". Being a group blog means that the main posts are written by several people — three in this case, John, David and Urs. The blog includes a short description of each of them. John Baez is a mathematician and a mathematical physicist, who specializes in quantum gravity and n-categories, and who is interested in many other areas as well. Gina did not know what n-categories were, but was looking forward to learning. She was very excited to discover that John Baez was a relative of the legendary singer Joan Baez. Urs Schreiber is a physicist and his work is related to string theory. And David Corfield was working on "the Philosophy of Statistical Learning Theory". Gina likes philosophy and greatly admires statistics.

Challenged by ksh95, Gina was determined to try to participate and make her mark on the n-category cafe!

John   David   Urs

17   From Perception of Mathematics to Dyscalculia

"I think that already commutativity of multiplication is subtler than commutativity of addition, so it's not so surprising that commutativity of exponentiation breaks down entirely"

Toby Bartels

David Corfield's post "knowledge of the reasoned fact" dealt with abstraction and causality in mathematics. David asked: "what to make of our expectation that behind different manifestations of an entity there is one base account, of which these manifestations are consequences?" He also referred to the idea that "mathematics has a causal/conceptual ordering".

David Corfield's two ideas can be explained with very simple examples. (Perhaps, they are too simple.) We start with the first idea: The fact that three apples plus five apples sum up to eight apples, and the fact that three oranges plus five oranges sum up to eight oranges both manifest the same "base account" which is "5 + 3 = 8". As for the second idea, perhaps this is an example: The fact that 8 − 3 = 5 comes as a consequence of the more basic fact that 5 + 3 = 8 in the causal/conceptual ordering of the truths of mathematics that David refers too.

David Corfield's own example for one "base account" with many different manifestations was the normal distribution in probability theory which is described by the "bell curve". The normal distribution and the bell curve occur in many different places.

Gina was very interested in children's perception of mathematics and wondered if it may be relevant to what David called the "base accounts" of mathematical concepts, and if the order in which children learn mathematical concepts has something to do with the "conceptual ordering" David was asking about. She was also curious as to how things she learnt in university fit into David's framework.

Posted by: Gina on October 24, 2006 4:19 AM

Re: Knowledge of the Reasoned Fact

Dear David,

It appears to me that for some aspects of the issues you raise it would make sense to consider

a) Very basic mathematical facts that we learn in school or early college

b) Naive understanding (by children/students) of mathematical phenomena, their reasoning and the causality relations between them

Perhaps we need something like Chomskian linguistics for mathematics.

Here are some examples:

Elementary school mathematics: (ask children, compare to what professionals say) what is the reason for

1. $5+8 = 8+5$

2. $6 \times 5 = 5 \times 6$

(Is 1. part of the reason for 2.?? is there a common reason??)

3. $42 - 19 = 23.$

High school:

4. Why a^b is *not* equal to b^a?

5. The sum of angles in a triangle is 180 degrees.

College:

6. The square root of 2 is not a rational number

7. A continuous function that takes a positive value at *a* and a negative at *b* takes 0 in between.

And a question that bothered me as a student, to which I have never been given a satisfying answer:

8. Why is it that a real function that has a derivative at any point may fail to have a second derivative, but a complex function, once having a derivative at every point, automatically has second and third and higher order derivatives??

Anyway, while talking about reasons and causality in mathematics, it may be good to compare insights of laymen and "professionals" and to consider concretely some examples like those suggested above.

This was certainly one of Gina's most successful remarks, and it led to many interesting discussions in various directions. The first three items on elementary arithmetic were especially popular and led to a separate discussion on "commutativity." The items related to college education have led to an interesting discussion on the notion of "continuous functions" and to an exciting description of the notion "elliptic differential equations." John Amstrong related to the idea of studying perception and learning of mathematics, and the structure of mathematics itself in a similar way to Chomsky's theory regarding language.

Posted by: John Armstrong on October 24, 2006 5:33 AM

That's a great idea, but in view of the reality of the situation I am pessimistic.

...

Chomskian linguistics works because everyone uses language and it seems perfectly natural to ask "why" questions about it. By and large most people are content to think of mathematics at even the most basic level as some sort of esoterica. The only reason any of it is true for most people is that that's what they were taught, if they remember it at all.

A long discussion about commutativity followed. Commutativity refers to a very simple yet important rule in mathematics. The commutativity law of addition asserts that $a + b = b + a$, and the commutativity rule for multiplication claims that $a \times b = b \times a$. In her remark Gina wondered how children perceive these rules, are they simply obvious? Are they obvious when it comes to one digit numbers, but less so with many digits?

Toby Bartels said: "I think that already commutativity of multiplication is subtler than commutativity of addition, so it's not so surprising that commutativity of exponentiation breaks down entirely."

Gina said: "I would conjecture that children usually understand why sum and product are commutative and the meaning of subtractions but the algorithms for arithmetic operations on 2-digit numbers obscure this understanding."

John Armstrong said: Children "understand" that $2+3 = 3+2$, because they can see it on their fingers and the notion of the order they put their fingers up in is out of mind. "How could it be any other way?" they think.

Toby Bartels proposed set theory as a tool to explain commutativity of addition and multiplication. **John Armstrong replied:** "I think these interpretations of the 'meaning' of arithmetic operations are interesting and definitely worthy of consideration. I also think that they move even further from the question, which is how students learning these arithmetic operations think about them."

It took Gina a few seconds to realize that "the question" John referred to was her own question which had become central to this discussion. "Wow" she thought to herself.

John continued: "Gina's point in raising those particular examples was to suggest a parallel to Chomskian linguistics: ask children learning the concepts "why" they are true, and that will give you insight into their epistemic causality. The idea is that the first stabs of an unsophisticated observer towards an explanation contain a deep insight into how the human mind processes the concepts. This is to be contrasted with the viewpoint of an expert who has already thought long and hard about the nature of the subject, and who cannot simply "unknow" that knowledge."

Next came David, who explained the difference between his own quest for an ordering of all mathematical statements based on causality, and Gina's interest in the perception of children. And John and John engaged in a little exchange of details about "new math". John Baez said: The New Math may not have helped everyone, but it helped me and John Armstrong concluded: "The New Math foundered on many shoals, but its heart was in the right place."

John Armstrong

Gina enjoyed the widespread discussion that followed, but children's perception of mathematics was closer to her heart.

Posted by: Gina on October 25, 2006 8:35 PM

Dear all,

I agree with John that a Chomskian study of children's learning, reasoning and insights about counting, arithmetic and mathematics will have a limited scope compared to a similar study pertaining to languages. It may, however, still be useful. Besides the philosophical issues of mathematical causality, it can be relevant to the understanding of **dyscalculia** — learning disabilities related to mathematics (Math Dyslexia).

Children's learning disabilities is a loaded and complicated subject, thought Gina. When it comes to mathematical abilities and disabilities, it appears that very little is known. The amount of Google hits for "dyscalculia" is two orders of magnitudes lower than for "dyslexia." This is a topic awaiting further study.

Chomskian Linguistics

The Chomskian revolution in linguistics is comprised of three elements. The first is finding common structures and formulating common rules that apply to all human languages (to a much greater extent than before). The second is relating linguistics to studying and making hypotheses about the way children acquire languages. And the third is studying mathematically very abstract forms of languages. Chomsky's theory of generative grammar is important in all three aspects.

Chomsky's perception and demonstration of the unifying concepts behind different languages have impacted the way languages are perceived by linguists and by philosophers, and dramatically changed the way linguistics is practiced. Most relevant to the thread above, Chomsky saw a direct link between the way children acquire language and the internal structure and logic of languages. His works in this direction are regarded as part of the cognitive revolution in psychology. While emphasizing the universal rules behind different grammars, Chomsky also made a strong point regarding the uniqueness of the cognitive aspects of language as compared with other cognitive abilities. He had a famous debate with psychologist Jean Piaget on this subject. Chomsky's mathematical works on formal languages and the related concept of "automaton" are now fundamental in theoretical computer science.

Chomsky is criticized for being too dominant in the area of linguistics and for leading to unmotivated sharp turns in his own theory. The decline of individual language studies is regarded by some as a negative side-effect of the Chomskian revolution. Others argue that without a major additional statistical ingredient, formal mathematical structures á la Chomsky's generative grammar and "transformation rules" are insufficient for understanding the structure and acquisition of languages. Some even claim that bringing major computational and statistical ingredients would make Chomsky's theory obsolete.

In English this is "anti anti anti anti missile missile missile missile missile"; in Hebrew (with words translated to English) it is "missile anti missile anti missile anti missile anti missile".

18   From Lehrer's Song
on Continuous Functions,
to Baez's Lyrics on Elliptic PDE

"In my youth I mainly did hyperbolic partial differential equations (PDE), since I liked physics and especially waves. I looked down on the elliptic folks for working on static phenomena like bubbles and taut drumheads — no life, just sitting there, perfectly smooth. But, elliptic PDE certainly have their charm,"

John Baez, October 26, 2006 1:58

The last items on Gina's list were about college education. One question was about "calculus" and the notion of continuous functions. "Continuity" is a notion you learn in calculus that is difficult to fully understand. This led to an interesting little thread devoted to continuity, and Gina was especially grateful to Blake Stacy who provided a link to a song by Tom Lehrer:

Blake Stacey, on October 24, 2006 3:10 PM

Gina asked about the following...:

7. A continuous function that takes a positive value at *a* and a negative at *b* takes 0 in between.

How can we explain what a "continuous function" is? A textbook might do so as follows: In everyday speech, a 'continuous' process is one that proceeds without gaps of interruptions or sudden changes. Roughly speaking, a function $y = f(x)$ is continuous if it displays similar behavior, that is, if a small change in x produces a small change in the corresponding value f(x). This statement is rather loose and intuitive, and intended more to explain than to define. To give a real definition, we break out the machinery of limits and begin employing deltas and epsilons, as Tom Lehrer describes here:

Tom Lehrer, "There's a Delta for Every Epsilon",

There's a delta for every epsilon,
It's a fact that you can always count upon.
There's a delta for every epsilon
And now and again,
There's also an N...

We come to the last item on Gina's list. Gina's last question was especially complicated. It was about functions over the complex numbers. The complex numbers are obtained by adding an "imaginary" number "i", the square root of -1. When you take the ordinary numbers (which are called "real numbers") and add 'i', and allow addition and multiplication, all sorts of miracles occur.

Gina asked about complex functions and about one of those miracles. In response, John Baez talked about "partial differential equations" (PDE in short). Studying PDE is important in mathematics and in most of its applications. John related the miracles of complex numbers and complex functions to a special family of PDE — elliptic PDE. Understanding John Baez's reply requires some college mathematics, but perhaps the beautiful Baez music can be appreciated even without understanding the lyrics.

Posted by: John Baez on October 26, 2006 1:58

Gina writes: Why is it that a real function that has a derivative at any point may fail to have a second derivative, but a complex function, once having a derivative at every point, automatically has second and third and higher order derivatives?? I would be happy to hear if there is a "reason" or "intuition" behind the miraculous difference between real functions which have a derivative at every point, and complex functions with the same property.

There certainly is a reason, and when I teach complex analysis I try to explain it.

After all, this is one of the biggest pleasant surprises in mathematics. In real analysis you have to pay extra for each derivative: for example, most functions that are 37 times differentiable do not have a 38th derivative. But in complex analysis being differentiable *once* ensures a function is *infinitely differentiable!* It's as if you bought a doughnut at a local diner and they promised you free meals for the rest of your life! And some people say there's no such thing as a free lunch....

So, we have to understand this seeming miracle.... The short explanation is that a differentiable complex function automatically obeys an elliptic differential equation, so it's as smooth as a bubble or drum-head!

John Baez gave a long detailed explanation and described also his personal feelings towards different classes of partial differential equations.

19   Foundations

In David Corfield's next post titled "Foundation" he asked: "Something that has long troubled me is the question of why philosophers have shown what, to my mind, is an unwarranted interest in 'foundational' mathematical theories which make little or no contact with mainstream mathematics." The case for category theory as a foundational mathematical theory was prominent in the thread.

In response to David, Gina said that she was always puzzled why philosophers were interested in mathematics in the first place. She also asked what category theory could offer philosophers and computer scientists.

Posted by: John Baez on October 30, 2006 7:38 PM

Gina asks why computer scientists should be interested in category theory. That's a good question. I should answer it sometime, in a nontechnical way. This thread is not the right place.

For now, I just want to note that computer scientists *are* interested in category theory. They don't need any convincing. There are lots of conferences on categories and computer science, just like there are lots of **books** and **review articles**. So, there's no need for any categorical proselytizing, at least in computer science. The train is already rolling. Here at the *n*-Category Café we can just **jump on**!

Why should computer scientists be interested in category theory? Gina was reminded of the following dialogue:

A: "Why should there be 'z' in 'money'?"

B: "But there isn't 'z' in 'money'!"

A: "So why isn't there a 'z' in 'money?'"

B: "Why should there be 'z' in 'money'?"

A: "That's what I asked you in the first place, why should there be 'z' in 'money'?

Kenny Easwaran explained why philosophers are interested in mathematics:

> Logic has clear philosophical relevance, as does our knowledge of logic.
>
> The case for basic arithmetic being of philosophical interest is only slightly trickier. The notion of ontological commitment is extremely important in analytic metaphysics. When I say "the number of planets is 8", there's a question of whether I'm just committing myself to some planets, or to a number as well.

Gina was drawn to what Kenny Easwaran wrote but, at the same time, she could not understand it at all. Kenny continued:

> Frege, and recently others, have wanted to show that this commitment actually already comes from logic and from some constitutive principles about the meaning of "number", thus suggesting that it isn't really a problem for metaphysics. Many philosophers are skeptical about whether areas of mathematics beyond logic and basic arithmetic are actually of general philosophical interest.

Can Category Theory Serve as the Foundation of Mathematics?

Usually the foundation of mathematics is thought of as having two pillars: mathematical logic and set theory. We briefly discussed mathematical logic and the foundation of mathematics in the story of Gödel, Brouwer, and Hilbert (Chapter 2). The story of set theory is one of the most exciting in the history of mathematics, and its main hero was George Cantor, who discovered that there are many types of "infinity."

Mathematical logic was always considered as a very abstract part of mathematical activity, related to philosophy and quite separate from applications of mathematics. With the advent of computers, however, this perception completely changed. Logic was the first, and for many years, the main mathematical discipline used in the development of computers, and to this day large parts of computer science can be regarded as "applied logic."

While mathematical logic and set theory indeed make up the language spanning all fields of mathematics, mathematicians rarely speak it. To borrow notions from computers, mathematical logic can be regarded as the "machine language" for mathematicians who usually use much higher languages and who do not worry about "compilation." (Compilation is the process of translating a high programming language into machine language.)

The story of Category Theory is markedly different from that of mathematical logic and set theory. It was invented to explain abstractly a certain area of mathematics called "algebraic topology." Specifically, the area of algebraic topology was based on a certain mathematical trick of associating algebraic objects to geometric objects, and category theory began by giving an abstract explanation to this trick and, along the way, to various other tricks from different areas of mathematics, which seemed unrelated. Category theory can be regarded as an abstraction of mathematicians' practices, even more than of mathematical notions. Amazingly, this abstraction of a single mathematical area turned out to be a very useful language, and a practical way of thinking in many (but not all) mathematical areas.

It is not easy to give a popular explanation of what categories are, because the notion is based on a familiarity with modern mathematics. One experience we have when we study mathematics is that the same method, or even the same equation, may solve very different problems. This seems to express the "abstract" power of mathematics. Within mathematics itself, we are often interested in knowing when two mathematical structures are essentially the same. The technical word that expresses the equivalence of two mathematical structures is "isomorphism." Category theory can be described as adding one more level of abstraction: trying to understand, in an abstract way, "isomorphism" and related notions.

Here are a few more details. Many mathematical areas can be described in terms of the "objects" studied in them, as well as in terms of certain notions of "maps" (functions) between the objects. (So in set theory the objects are "sets" and the maps are "functions," and in group theory the objects are "groups" and the maps are called "homomorphisms.") Categories are mathematical gadgets that put these common structures on common abstract ground. Every category has "objects" and "morphisms" with some abstract properties. An important notion is that of a "functor", which is a way to relate one category to another. Categories indeed seem to play a pivotal role in the foundation of mathematics, or at least in some of its major areas, but they constitute a different sort of foundation. If we compare logic and set theory to the "machine language" of computers, we can regard category theory as an extremely useful universal programming tool.

Even if you did not follow the details about category theory, perhaps you got the correct impression that category theory is part of the mathematical trend to make things more and more abstract. Is this a good trend? Some opponents of category theory refer to it as "abstract nonsense." But even they concede that it is sometimes extremely powerful. There is a healthy tension in mathematics between the ongoing efforts to understand things more abstractly and the efforts to understand more and more concrete issues and examples.

Samuel Eilenberg (left) and Saunders Mac Lane who introduced category theory.

Following is an ad-hoc discussion about category theory as a foundation for mathematics between Menachem Magidor (a set theorist and, at that time, the president of the Hebrew University of Jerusalem), Azriel Levy (a set theorist and Magidor's thesis advisor), and the editor of this book. The conversation takes place in the main corridor of the Mathematics Department, where Menachem is standing and smiling the smile of a person visiting his beloved hometown after a long absence.

Menachem Magidor

GK (surprised): Hi Menachem, what are you doing here?

MM (tries unsuccessfully to look offended): What do you mean? Am I not wanted here?

GK: No, no, no, no, I was just wondering if the state of the Hebrew University is so good that you can afford to visit us.

MM: No, on the contrary, the situation is hopeless... Seriously, I was just taking a book...

AL: (enters the corridor, surprised) Shalom Menachem, what are you doing here?

MM (once again tries unsuccessfully to look offended): What do you mean? Am I not wanted?

AL: No, no, I'm just surprised that you have the time...

GK: (interrupts) Guys, I have a quick question for both of you, can category theory serve as the foundation of mathematics instead of set theory?

MM: (eagerly) Hmm, it is actually an interesting question, there is a result that a certain topos theory has the same power as a set theory with certain axioms ... of course you need some separation and replacement axioms ... do you remember, Azriel?

AL: The crucial question in my mind is that sets are very easy to explain. Can you explain topos theory to high school students?

Other people arrive and the discussion is interrupted.

20   Computers, Categories, Analogies, and Greatness

"Good mathematicians see analogies. Great mathematicians see analogies between analogies."

Attributed to Stephan Banach

The following post a few days later was a pleasant surprise:

A Categorical Manifesto

Posted by John Baez

A while back **Gina** asked why computer scientists should be interested in categories. Maybe you categorical computer scientists out there have your own favorite answers to this? I'd be glad to hear them. To get you going, here's one man's answer:

Joseph Goguen, A **categorical manifesto**, *Mathematical Structures in this Computer Science* **1** (1991), 49–67.

Gina was of course very flattered that her comment started a whole thread. The only remark which really dealt with computer science was by Dan Piponi, who explained some of the connections between category theory and computer science. The subsequent remarks dealt with category theory and not with computer science. But then Jonathan Vos Post put a nice twist into the discussion.

Posted by: Jonathan Vos Post on November 11, 2006 3:09 AM

Ulam on Banach; Re: A Categorical Manifesto

In his autobiography, Stanislaw Marcin Ulam, a mathematician of the Lvov School of Mathematics, attributes the following to Banach:

"Good mathematicians see analogies. Great mathematicians see analogies between analogies."

Analogies

Vos post continued to discuss Banach's statement but Gina remained skeptical and was in the mood for a good debate:

Posted by: Gina on November 13, 2006 1:23 PM

Banach (according to Ulam) said: "Good mathematicians see analogies. Great mathematicians see analogies between analogies." And Jonathan asked if we can even say that 'very great mathematicians see analogies between analogies between analogies between analogies'?"

Actually, Banach's saying sounds appealing but rather suspicious, as shown by Jonathan's suggestion that seems to push it towards the absurd.

We can check this with some examples: I am sure there are many examples of useful analogies in mathematics (what are the best ones, in your opinion?) but can you give a few examples of analogies between analogies? And can you give a single example of an analogy between analogies between analogies??

Does Banach's saying (or Jonathan's) make sense if we replace "mathematician" with "scientists","police detectives","medical doctors"??

So when scrutinized, Banach's statement may not be that much better than the common public perception:

"Good mathematicians can multiply 2 digit numbers in their heads and great mathematicians can multiply 5-digit numbers in their heads!"

The discussion that followed was fairly technical. John Amstrong mentioned the analogy between spaces and groups and the analogy between fields and symmetries, and then noted the analogy between these two analogies that has to do with a mathematical notion called "adjointness". This looked like a very good example, but Gina wondered if it supported Banach's statement. Who were the good mathematicians behind John's two analogies, she asked. Those were Poincaré and Galois, and Gina wondered who could be the great mathematician behind the analogy between the two analogies. Then more examples followed and at the end, Gina was somewhat convinced (or tired) and said: "Banach could have just said: 'among various other things, good and great mathematicians spend their sweet lives by seeking analogies and analogies between analogies.'

"But apropos Banach's saying," continued Gina, "what about 'greatness': is it **the** moving force behind it all? Let's discuss Greatness and the pursuit of greatness."

Greatness is an intriguing subject to pursue, thought Gina. But this topic was not picked up.

It was a good time to bid farewell to the *n*-category café as Gina wanted to return to the string theory debate. This time, it was a friendly goodbye.

Part III

Asymptotia and Lee Smolin's "The Trouble with Physics"

Clifford Johnson

"Asymptotia" was the weblog of string theorist Clifford Johnson. Clifford ran a series of posts titled "Tales about storms in a tea-cup I, II, III, IV, V and VI" (and later VII) dealing with the string theory controversy and, in particular, with Peter Woit's and Lee Smolin's positions and books.

Clifford was Gina's favorite blogger. He seemed relaxed and let the discussion flow. He did not behave like an ultimate authority and, in fact, did not interfere too much. Gina did not agree with Clifford's overall approach towards the debate. She wrote: "Clifford's name of the series of posts on this controversy "More scenes from the storm in a teacup" may seem demeaning, but it need not be, 'intellectual storms in teacups' may very well be an important part of what academia and science are about."

21 Can Strings be Fractals? and All Sorts of Other Things

Debates and controversies

Some people complained about the cover of Lee Smolin's book and found it too insulting. Others considered the analogy with the naked emperor insulting. On the other side of the debate some people even suggested that negative book reviews are unfair since they cause people not to buy books they would otherwise buy. Gina strongly disagreed.

73 — Gina Nov 12th, 2006 at 4:44 pm

There is nothing wrong with the cover of Smolin's book, in fact it is nice. (There was also nothing wrong with Lubos Motl's Amazon review of Smolin's book.) There is nothing wrong with the naked emperor metaphor either. And there is nothing wrong with the provocative name "not even wrong", in fact it is also nice. Exact-science guys are simply not used to practicing and maintaining debates and controversies. (They are either being extremely puritan or occasionally unreasonably aggressive.) Try to consult your colleagues from the Humanities and Social Sciences!

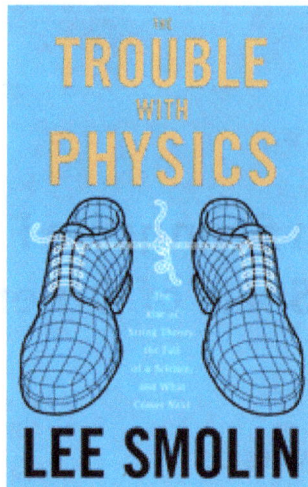

The cover of Lee Smolin's book

The Hype

90 — Gina Nov 18th, 2006 at 9:13 am

It was claimed in these discussions that string theory is over-hyped, and perhaps it is. But let me argue that this hype served not only string theory but also, to a large extent, physics, mathematics, and science as a whole. Society's exposure to sometimes bold (but sincere and serious) claims about science can serve science. On the other hand, hype cannot change even a little bit the factual scientific reality or ease even slightly the immense difficulty in exploring it.

Can Strings be Fractals?

Gina Jan 21st, 2007 at 8:03 pm

Dear Clifford,

I have the following question about strings. (Probably it puzzles others as well.) From what I saw on Wikipedia strings are like arcs that can close up or remain open, and if you add time, you get a nice and smooth two dimensional picture like a cylinder. Is this the right mental picture to have in mind?

My question is, why one-dimensional and not, say, 1.25-dimensional? Do you regard string as a nice looking curve that over time looks like a nice surface? (Or it is just the pictures.) I remember that there are curves that look very different: like the famous Peano plane filling curves. These curves have dimension 2 and there are even curves that have a dimension which is not an integer number. So isn't it better to assume strings are some sort of fractals? Say they are 1.25-dimensional (or 1.63-dimensional) arcs in space? Does string theory rely (beyond the nice pictures) on thinking about strings as smooth and nice, and their "spacesheets" as nice behaving surfaces? Does it make a difference if strings are fractals?

Fractals

Fractals are beautiful mathematical objects, the study of which goes back to the late 19th century. The Sierpiński triangle and the Koch snowflake are early examples of fractals which are constructed by simple recursive rules.

Koch snowflake (left) and Sierpiński triangle (right). (Note: The Koch snowflake is just the "boundary" of the blue shape in the picture.)

Other examples are based on the study iterations of simple functions, especially functions defined over the complex numbers.

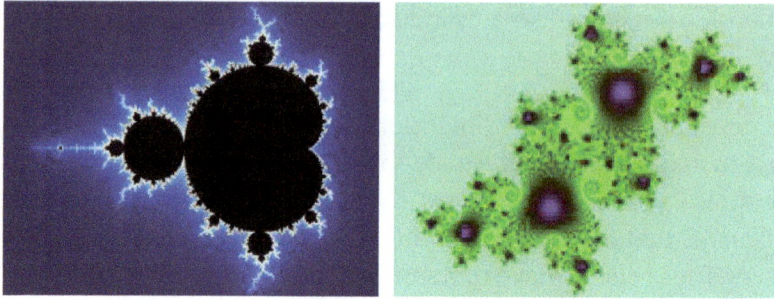

Mandelbrot set (left) and Julia set (right).

Other examples come from various stochastic (random) processes. For instance, the outer boundary of a Brownian motion in the plane, and the boundary of the percolation process (random Hex game) that we considered in Chapter 9.

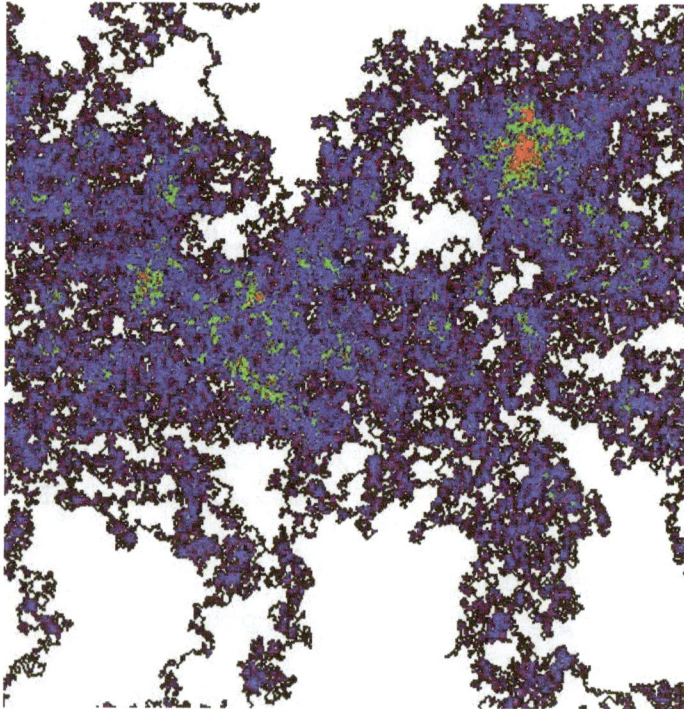

Brownian motion in the plane. The boundary is the "border" between the white areas and the colored areas. (If you get the impression that "boundary" is an important notion in many areas you are correct!)

We already mentioned the importance of the notion of "dimension" in mathematics. A point has dimension 0, a line has dimension 1, the plane has dimension 2 and space is 3 dimensional. Fractals often have "fractional dimension." The Koch snowflake has dimension 1.2619, and the Sierpiński triangle has dimension 1.5850. The "boundary" of the Brownian motion in the plane is a fractal; Mandelbrot conjectured that its dimension is 4/3 and this was recently proved by Lawler, Schramm, and Werner. The (marked) border between the white and grey areas in the percolation picture of Chapter 9 also leads to an interesting fractal.

The term fractal was coined by Benoît Mandelbrot, who in his book also proposed the following informal definition of a fractal: "A rough or fragmented geometric shape that can be subdivided into parts, each of which is (at least approximately) a reduced-size copy of the whole." An important property of fractals is referred to as "self-similarity", whereby a small part of the big picture is very similar to the whole picture.

Mandelbrot also understood and promoted the importance of fractals in various areas of physics. Indeed, today fractals play an important role in many areas of modern physics (and there is also some controversy regarding their role). Mandelbrot also wrote a paper concerning applications of fractals in finance.

The notion of self-similarity is also important in other areas. In computer science, the self similarity of a problem is referred to as "self-reducibility", and this property facilitates the design of efficient algorithms for solving the problem.

A fractal-shaped cauliflower, photographed, presented, (and eaten(?)) by Clifford Johnson.

A quilt by Anna-Maria Brenti, which is based on several mathematical constructions. In the bottom-right corner we see the "Menger–Sierpiński sponge," a fractal of dimension 2.7268 constructed by taking away pieces of the unit cube.

22   The Trouble with Physics: Gina's Sixteen Objections

At some point down the discussion, Lee Smolin complained that people hardly relate to, and try to argue with what he actually wrote in his book. Gina found this complaint rather justified, and also an opportunity for discussion on some specific claims made by Lee. [Editor's note: apparently Gina referred to sixteen as "baker's 15."]

139 — Gina Nov 30th, 2006 at 4:45 am

[Lee Smolin:] "I can count on one hand the number of times someone has made in print or online an argument of the sort, 'Smolin, or Penrose or ... makes the following argument…..leading to the following conclusion…., but it is wrong because he makes the following mistake in reasoning…..'"

Here are a baker's 15 points, some central, which, in my opinion, point to the incorrect, or, at least, problematic aspects of Lee's argument. (Dear Lee, I truly think that trying to attack your argument can be of service. There are, of course, things I like in the book.)

I will be happy to elaborate on each of these points.

1. Asserting that there was no real progress on the five major problems of physics described in Chapter 1 of Smolin's book is incorrect, or at least very problematic.

2. (Even ignoring 1.) Interpretation of the lack of progress, or the limited progress in answering the five major problems as a definite failure is very problematic.

3. The interpretation of the "Mandelstam story" as mainly damaging to string theory is incorrect.

4. The interpretation of Maldacena's conjecture is biased. Lee's interpretation is not sufficiently positive towards the ST endeavor, even unreasonably so.

5. The substantial reliance of Lee's case on how to interpret Maldacena's conjecture is in itself damaging to his case.

6. The issue of pluralism should be studied from a wider perspective.

7. Lee overlooks (or neglects) a more serious concern about pluralism in the high energy physics community (which is potentially problematic in terms of his other claims, however.)

8. It is strongly implied by Lee, but not explained, that it is not reasonable that string theory and drastically different quantum gravity approaches should develop separately, e.g., that string theorists should be on top of development in LQG (loop quantum gravity). As a matter of fact, it seems quite reasonable that (,e.g.,) ST and LQG would develop separately.

9. Contrary to Lee's suggestion, young scientists should not be encouraged to attack fundamental physics problems head on.

10. The negatively portrayed reactions to sporadic bold theories and ideas (like 't Hooft's) are, in fact, very reasonable. The current attitude of scientists is essentially the correct attitude.

11. Some natural, often welcome social processes in academia (regarding academic judgment and decision making) are depicted negatively. This is a mistake.

12. At a closer look, many claims, stories and quotes against the string theory community are not really damaging to the ST community, and mentioning them weakens the quality of Lee's argument.

13. Simultaneously talking about scientific and social issues, including funding, is in itself very problematic. Not being sufficiently aware of this difficulty is a mistake.

14. Although a substantial part of Lee's argument is about sociology, there appear to be no real interactions with social scientists.

15. Lee's ideas on revolutionizing universities are not detailed, and the hints we get are not promising.

16. Lee claims that science clearly calls for a revolution, as it is now more conservative (anti-revolutions) than ever. This is incorrect. In fact, maybe the opposite is true. The whole "revolution" terminology is off-base.

140 — Lee Smolin Nov 30th, 2006 at 8:00 pm

Dear Gina,

Thanks very much. I'd be very interested if you expanded some of these points. To make it easier to manage, shall we take one at a time? If you elaborate on one and explain why you disagree I will respond accordingly.

Thanks,

Lee

"I am in business", thought Gina, and was at the same time a little terrified as well.

Physics: Computations and Words

The issue of the mathematical rigor of physics computations, notions, and results is of great importance. Lee Smolin criticized the lack of rigorous proofs for certain fundamental results in string theory, and also raised another important point, which is almost in the opposite direction. He argued that the heavy "computation-based" methodology of particle physics, which was responsible for the great successes of the "standard model", is no longer suitable for further progress. Smolin sees more room for philosophical arguments and verbal physics ideas. (Smolin considers his approach to be in the tradition of the great physicists of the first half of the twentieth century.) Other physicists strongly disagree. Clifford Johnson regards "verbal physics arguments", which are not supported by computation, as "popular science." In the blog discussions physicist Joe Polchinski wrote that he regards the transition from verbal description to mathematical formalism as the step in which most of physics ideas fail. Joe also explained how his own career was transformed when he relaxed his standards for mathematical rigor, and wrote: "I used to focus too much on rigor and formalism, and have become a much more creative and productive scientist since learning, very slowly, to see through these to the physics."

A Brief Description of Smolin's Book

Smolin's book has four parts. The first part discusses physics, its main insights and achievements as well as its main research problems. It begins by stating the five great problems in theoretical physics, and goes on to discuss the beauty of physical theories and the "beauty myth", the role of geometry, and the importance of "unification", which brings together different theories under the same roof. Finally, Smolin reaches the question of quantum gravity — the unification of quantum mechanics and Einstein's theory of gravitation.

The second part describes string theory, which (among other things) is a theory of quantum gravity. Smolin describes the first revolution in string theory — the realization of the important role of supersymmetry, and the second revolution — the emergence of a new "M-theory" that unified earlier different string theories. He describes the string theory landscape (that Smolin himself was among the first scientists to discover), and discusses the anthropic principle, an attempt to deal with the difficulties posed by the landscape.

The third part describes speculative physics not directly related to string theory. Smolin skeptically examines many truths from physics and offers some radical alternatives and ideas. He mentions several alternatives for quantum gravity, including "Loop quantum gravity", which is a theory he was among the first to introduce, "twistor theory" which is mainly associated with physicist Roger Penrose, and "non commutative geometry" developed by mathematician Alain Connes. Smolin raises questions concerning the foundation of quantum mechanics, and mentions a theory called MOND introduced by physicist Mordechai Milgrom. This theory suggests that Newton's law of gravitation breaks down for highly accelerating bodies.

The fourth part of the book is about sociology in science and academic life, and about fundamental issues regarding the question of "what is science." Most of Gina's points are related to this final part, which Smolin regards as the most important contribution of his book.

23   TTWP: Mandelstam; Optimism

"Give me a couple of years free from other duties, and I shall complete the task — the task of copying the whole Book of Nature, and of writing the new science."

Francis Bacon, according to Carl Popper.

Francis Bacon

The story about "finiteness of string theory" and the results in this direction by physicist Stanley Mandelstam play an important role in Smolin's argument and occupy a very large percentage of the debate. String theory replaces point particles with curves and studies them using certain complicated infinite sums. Each individual term in those infinite sums is itself the outcome of a complicated mathematical process. Miraculously, at the right dimension, these terms become meaningful. This is the content of Mandelstam's effort and results. Gina felt that, although this issue was certainly important, it had gradually become a "red herring." Too much was made of this particular point.

143 — Gina Dec 1st, 2006 at 11:56 am

Thank you very much, Lee, for your willingness to discuss this matter with me, and thank you very much Clifford for your willingness to host it in the back yard of this old post.

As you suggested, Lee, let's take it one at a time. I will elaborate on each one of my points in my free time, you are most welcome to reply, as you see fit. Unless emergency occurs, I will not reply to your replies; here, we do need to converge. Points 3–5 are furthest away from my own interests in this debate so let's start with them. My third point was:

3. The interpretation of the" Mandelstam story" as mainly damaging to string theory is incorrect.

Here we refer to the story about the "finiteness of string theory". Important early evidence that string theory is going in the right track was that certain mathematical difficulties (called "anomalities") were resolved for the models proposed in string theory. Some of this evidence was based on the work of physicist Stanley Mandelstam. (Some evidence was found even before Mandelstam.) As Lee argues in his book, Mandelstam did not present a proof but rather partial mathematical evidence and it took, according to Lee, 17 (!) years until further stronger mathematical evidence was presented.

This story is overall not damaging to string theory but rather obliging. Mandelstam's story is cited by Lee in the context of string theorists' optimism (or excessive optimism). The meaning and nature of optimism in science is probably a topic for a whole separate discussion. Here we have a case where, as far as I can judge, both the overall optimism and the mode of operation of "moving on" were justified.

144 — Peter Woit Dec 1st, 2006 at 3 :09 pm

Gina,

It seems to me you completely mischaracterize and misunderstand Smolin's argument here. Why do you think it is a good idea to publicly criticize Smolin over an issue you admit that you don't understand?

Smolin was not saying string theorists should not have "moved on" and conducted research under the assumption that the conjectured multi-loop finiteness would work out, he was just saying that he was surprised to find that almost all string theorists he asked thought that this conjecture had been proven by Mandelstam, when it hadn't been.

Gina was quite surprised by Peter Woit's comment. On his own weblog he accused Gina of wasting his time and here he volunteered to respond to a question addressed to Lee.

Peter continued and described another example of what he referred to as "this phenomenon:" Peter was very excited by results that indicated that a certain theory called "N=8 supergravity" is finite, although string theorists conjectured all along that this theory is not finite. (Gina thought that Peter is overexcited about these interesting developments, and that having conflicting and confusing pieces of information is a very common phenomenon when new scientific theories emerge.)

146 — Gina Dec 1st, 2006 at 4:38 pm

Hi Peter,

I raised 16 points where I felt Lee's arguments were either not correct or problematic. This is an academic discussion and not public criticism, and I truly think that such critique can be useful, even if I am wrong on **all** 16 points.

Three of my 16 points centered on more technical issues, but I feel that I can understand Lee's logical argument even without understanding the precise technical nature of "finiteness of string theory". I think that my interpretation of this issue is reasonable and my critique stands.

This was an example which was supposed to have shown that string theorists were over-optimistic. But in this case the optimism appears to have prevailed. A much better example would have been a central case where string theorists claimed one thing and eventually the opposite was proven.

Gina thought that Peter Woit's comment reveals a certain flaw in his approach to science and continued:

When you talk about "this phenomenon" it seems that you have a certain difficulty in the understanding of how science is practiced. At the end (in the best of cases) all pieces fit together and it all looks wonderful. But along the way there are many misunderstandings and conflicting pieces of information, and sometimes even mistakes.

Lee's reply was more to the point and quite interesting. He uses the term "perturbative finiteness" to describe Mandelstam's disputed result.

149 — Lee Smolin Dec 2nd, 2006 at 3:17 am

Dear Gina,

Thanks. The issue has a scientific component and a sociological component. Science first:

-Perturbative finiteness is a major element of the claim of string theory as a potential theory of nature. If it is not true then the case for string theory being a theory of nature would not be very strong.

-Perturbative finiteness has not been proven. There is evidence for it, but that evidence is partial. There is a complete proof only for genus two, which is the second non-trivial term in an infinite power series, each term of which has to be finite. The obstacles to a complete proof are technical and formidable; otherwise we would certainly have either a proof or a counterexample by now.

This is not an issue of theoretical physicists' rigor vs. mathematical rigor. There is no proof at either level. There is an intuitive argument, but that is far from persuasive.

Is string theory in fact perturbatively finite? Many experts think so. I'm afraid that if there were a clear way to a proof it would have already been found and published, so I find it difficult to have a strong expectation, either way, on this issue.

The sociological issue is not that most string theorists chose to continue working on other problems in the hope that someone would eventually resolve it. This is fine, we all do what we can do, and many of us theoretical physicists do not have the inclination or talent to work on such a difficult and subtle mathematical problem.

The problem is that many presentations of string theory, for the public as well as for colleagues, seem to have been misleading on this issue. Many people I spoke to were under the impression that perturbative finiteness was an established fact. Most review papers for physicists and popular books gave the impression that perturbative finiteness of string theory is a fact. Only a few characterize the situation correctly.

Was this a problem? I think it was. I don't believe anyone was deliberately dishonest about this, but there was an atmosphere in which belief in the theory went beyond the evidence in the published papers. Was this conducive to the fact that many people, both in and out of physics, were under the impression that the problem of quantum gravity was already solved by string theory? I am not an historian or sociologist, but I worry that this might have been the case.

You claim that my discussion of this issue is "not damaging to string theory". I don't think it is, if by "string theory" you mean that string theory is of interest as one of several interesting approaches. It is damaging to the claim "string theory" is "the only game in town", as some proponents continue to insist.

Thanks,

Lee

Optimism: Beat This I

Here is a quote from Karl Popper's paper "Science, Problems, Aims, Responsibilities" about Francis Bacon:

"According to Bacon, nature, like God, was present in all things, from the greatest to the least. And it was the aim or the task of the new science of nature to determine the nature of all things, or, as he sometimes said, the essence of all things. This was possible because the book of nature was an open book. All that was needed was to approach the Goddess of Nature with a pure mind, free of prejudices, and she would readily yield her secrets. Give me a couple of years free from other duties, Bacon somewhat unguardedly exclaimed in a moment of enthusiasm, and I shall complete the task — the task of copying the whole Book of Nature, and of writing the new science."

Optimism: Beat This II

Here is a tale from Arundhati Roy's book "The God of Small Things". In the book Margaret Kochamma tells the following joke to Chacko in Oxford, England, where the two meet: A man had twin sons... Pete and Stuart. Pete was an Optimist and Stuart was a Pessimist... On their thirteenth birthday their father gave Stuart — the Pessimist — an expensive watch, a carpentry set and a bicycle... And Pete's — the Optimist's — room he filled with horse dung... When Stuart opened his present he grumbled all morning. He hadn't wanted a carpentry set, he didn't like the watch and the bicycle had the wrong kind of tyres... When the father went to Pete's — the Optimist's — room, he couldn't see Pete, but he could hear the sound of frantic shoveling and heavy breathing. Horse dung was flying all over the room... "What in heaven's name are you doing?" the father shouted to Pete...

A voice came from deep inside the dung. "Well, Father," Pete said, "if there's so much shit around, there has to be a pony somewhere." (*The God of Small Things*, Flamingo, London 1997, p. 243).

Francis Bacon (left) and Arundhati Roy

Stanley Mandelstam, on the occasion of the 1991 Dirac Medal Award Ceremony

24 How to Measure Progress?

The next item on Gina's list was Smolin's view of Maldacena's conjectures. This was a very mysterious topic! Before embarking on that topic, however, Gina wanted to make sure she understood the gist of Smolin's criticism, and so wrote:

> My basic understanding of Lee's book is that his argument asserts the failure of string theory to reach any progress in the basic questions of physics in the last three decades. If Lee's argument regarding string theory (as a scientific theory) is just that string theory has not succeeded **yet**, or that there are some concerns about specific results, as well as about the whole endeavor, then this is a different matter. I do not think what Lee writes in his book supports such a mild interpretation of his stance.

Lee referred to Gina's question regarding his overall argument. No, he is not claiming that string theory has failed. This was most interesting!

To Gina

You wrote: "If Lee's argument regarding string theory (as a scientific theory) is just that string theory has not succeeded **yet**, or that there are some concerns about specific results, as well as about the whole endeavor, then this is a different matter." Yes, of course, this is the basic premise of my argument.

The conclusion I draw from this is that since there are other very well-motivated directions of which the same could be said, string theory should be considered and treated as one of a number of important approaches to quantum gravity and unification that are pursued. I also argue that it would be best for the progress of science if all those who work on these different approaches consider themselves as a single research community, within which we try to mix people doing different things, as well as vary our own research interests, because presently different questions are best approached with different approaches.

My major argument is not "against" string theory; it is against the idea that string theory is "the only game in town" or that it is the "modern paradigm," suggesting that it alone deserves intensive investigation within a community specializing in working on it alone.

Interesting, thought Gina, but she was not convinced that Lee's strong rhetoric, of "the rise of string theory and the fall of science" supports such a mild interpretation of his stance. Next, Gina related to another general issue central to the debate. How do we measure progress?

One issue that is relevant to all the previous items and to the whole discussion is the matter of "how to measure progress" and "how to measure partial success." This is an important question here, as in other areas of science (and life). If we are too quick to declare progress, we may end up nowhere, or with self-illusion and self-deception. If we are too slow, we may not even get started. I suggest keeping in mind the issue of "how to measure progress" for the discussion of other points of disagreement.

In response, Lee expressed his view on the question of evaluating the partial success of scientific theories.

To Gina,

I agree that the key question is what to do about research programs where we have presently partial but not total success. I suspect that what is confusing you is that my view of what is wise in these circumstances does have a tension in it. I do think that we must at all times be aware of the high risk these research programs entail, and therefore be prepared that partial success may be the best that we can accomplish. We live in an unfortunate period in which there are many research programs in theoretical physics that, in spite of being well motivated, achieve only some partial success but never go beyond that. So I believe that individuals should, of course, continue to develop programs that are partially successful, but that we should do everything we can to leave room for someone to invent a new program that may be more successful.

Thanks,

Lee

Answer: Lord Kelvin, the Age of the Earth, and the Age of the Sun

Lord Kelvin dating the Sun and Earth

The physicist William Thomson (who later became Lord Kelvin) spent a lot of effort in estimating the age of the earth. His estimations were between 20 million and 400 million years. We asked what was the main reason for Lord Kelvin's wrong estimation. Was it radioactivity, heat produced by radioactive decay? This was a process that would remain unknown to science for decades to come. Or was it convection — the transfer of heat not through radiation or heat-conduction but through the movement of hot parts to the surface?

The short answer is that the main mistake in Lord Kelvin's estimates was convection. This accounts for more than 90% of his error.

There is more to be said, of course. Kelvin (previously Thomson) was certainly one of the greatest physicists and engineers of the 19th century. He directed the laying of the first transatlantic telegraph cable, formulated the second law of thermodynamics and wrote hundreds of influential papers. At a very young age he was one of the first to study and understand Fourier's heat-flow treatment, and it is this that led him to calculate first the **age of the Sun**, then that of the Earth. Whereas the former calculation obviously did not take production by nuclear reactions into consideration, the latter calculation neglected to take convection into account.

Kelvin assumed that the source of the Sun's energy was gravitational — a collapse (accumulation) of many meteors, and thus conversion of mechanical energy into heat (Joule's heat equivalent was already known). Accordingly, he first calculated the Sun's age to be around 32 millions years, later putting an upper limit of 300 million years and a "most probable age of 100 million years." Kelvin (correctly) assumed that the Earth cannot be older than the Sun, and is most likely of the same age. The prevalent hypothesis for the formation of the Sun and planets was that of Kant-Laplace, which involved the above-mentioned "nebular hypothesis" (an improved version of it is still accepted today). So when Kelvin found the "age of the Earth" to be within error of what he assumed to be the "age of the Sun" both reached through different approaches, he was very happy.

It is rather impressive that this genius of physics was opposed by a bunch of geologists (high boots, shorts, field hammers, limited perspective) and that they were, in fact, right, but it is also true that Kelvin showed two important points:

The first is that the age of the Earth is NOT INFINITE as many thought in the 19th century, and the second is that the age of the Earth is calculable from physical principles. At that time, several geologists claimed that "physics cannot be applied to geology," and in this they were wrong! We owe him a great deal for teaching us these two points.

What an exciting, spirit-lifting story it is! (I heard it from my colleague Yeshu Kolodny.) And there are more facets to it. Of course, radioactivity played a major role in modern estimates for the age of the Earth.

25   Maldacena

"My own reaction to this was basically: 'there ought to be a different way around this'"

Roger Penrose [explaining his reaction to high dimensions required by string theory], The Road to Reality.

"You have this solid octopus and you hit it on its head and hear (wonderful) music, say, a Mozart sonata…"

> Gina, [trying to explain the ADS/CFT
> correspondence and the holography principle
> without understanding it,] Dec 2, 2006.

Juan Maldacena's conjectures and results are considered as a crowning achievement of string theory. Yet the content of these results, which are referred to as "the ADS/CFT correspondence" and the "holography principle," is very difficult to understand. (If this helps a little, ADS stands for "anti de-Sitter" and CFT stands for "conformal field theory.") Lee Smolin tried to find some weaknesses in the picture portrayed by Maldacena's work and the huge subsequent literature. As a scientific "devil's advocacy", such attempts (with merit) can be of interest. But to make these ideas part of a case against string theory was mind boggling. Gina could not understand Lee's overall argument regarding Maldacena's work.

160 — Gina Dec 2nd, 2006 at 4:42 pm

Dear Lee, Thank you for your reply and good spirit. I will elaborate on my 4th and 5th comments that are both related to Maldacena's conjecture, and will take the simpler and more important 5th point first. (If you find it fit to reply please indicate the number of the comment.)

5. The substantial reliance of Lee's case on how to interpret Maldacena's conjecture is by itself damaging to his case.

The fact that such a major part of Lee's argument deals with the validity and interpretation of Maldacena's conjecture speaks for the viability of string theory. Maldacena's conjecture is not a 19th century conjecture that defied proofs for centuries, but something that happened less than 10 years ago. Lee's argument against string theory without Maldacena's works around would have been stronger.

The issue of proofs and precise interpretation of Maldacena's conjecture is secondary to the issue of whether string theory is viable enough to produce developments of such magnitude in the future. I do not see any reason why we cannot expect such further development.

Will there soon be some new developments in string theory of the same magnitude as Maldacena's works (or bigger)? Nobody could know. And Gina considered this uncertainty to be an appealing feature of science. Gina took some more time to list her specific points regarding Maldacena's conjecture and then added a remark about how she perceives, in her laywoman eyes, Maldacena's discoveries.

Following Peter Woit's criticism of my first item, I did try to learn more about the ADS/CFT issue and looked at Penrose's book, "The Road to Reality". I did not make much progress but this was a delightful experience. Penrose discusses the technical details of string theory with affection and even enthusiasm, like talking about a beloved child, even if he does not subscribe to the theory at all. His skepticism is also charming, and he is also skeptical about his own skepticism. For example, look at what Penrose says about the high dimension idea — "My own reaction to this was basically: 'there ought to be a different way around this'" — how simple, how nice! You do not need more, and it is such a good line for many purposes — "There ought to be a different way around this". My heart goes to Roger! (And according to the pictures in Penrose's book the universe is simply a sort of amoeba or an octopus.)

Reading Penrose on top of what I already learned from Lee and other popular sources, I developed the following mental picture on this ADS/CFT: You have this solid Octopus and you hit it on its head and hear (wonderful) music, say a Mozart sonata. Now, you take just the very thin external skin (called the boundary) and you hit that. And you hear precisely the same wonderful music. Precisely the same!

169 — Lee Smolin Dec 7th, 2006 at 6:08 am

Hi Everyone

Sorry to have fallen behind. Here are some replies to recent posts: With respect to Maldacena, I don't want to repeat the arguments in my book and papers on this, but I want to make it clear that I believe that at whatever level, it is true the ADS/CFT conjecture is a major result in mathematical physics. But I do not think it is a good thing to treat the strong form of the conjecture as essentially true, given that it posits the existence of an isomorphism between two structures which have not even been written down.

No Witches in Portugal, by Michal Linial (Taxi Story no. 2)

I landed in Lisboa, Portugal last night. I need to get to my hotel in Cascais. The nice lady in the tourist information ordered a taxi and happily said: "Do not worry, our drivers speak excellent English."

Five minutes later my driver arrived, a huge person, nicely dressed with a nice fancy black car. "To Cascais, what is the price?" I asked in a fluent English. Well, it depends he replied…

My huge driver, was very fluent in English as promised… Then he started… "Three years ago I had a ski accident, at that time I was much heavier…." I was wondering how such information might be relevant to our trip… My driver continues: "From the time of the accident my leg is not the same. This is why I bought this car with a "gear" instead of an automatic transmission, so I can practice my left leg. Most of time I can move it, but I still do not feel it when it is cold"… "Is it cold today?" I asked. No, today it is fine. Yesterday it was cold.

How long is it to Cascais? I inquired. My driver replied honestly: "it depends on the road"… Oh, I said… waiting for an explanation. It took a minute for him to explain: "You know, I always drive by my feelings. Sometime, I have a feeling that it is not good to take the freeway. Today I have this feeling. "OK, it is best that you listen to your feelings", I replied quietly. He continued: "You know, there are no witches in Portugal, but it is better to listen to them". Of course, I said.

"You know" he continued, "my brother in law who was a driver in our taxi company had a bad feeling 4 months ago when he took the small road instead of the highway, and indeed he died. A truck crushed him. But the passenger, a British person that was in the car was not hurt. The passenger had his seat belt fastened, like you… Do you want me to show you where he died? It is on the way to the hotel," he insisted. "Sure'" I replied. "If it is on the way, and you feel that we should not take the highway, let's go on the other road…"

We headed down a neglected road. It was not too cold… My huge driver showed me the spot where his brother in law had died… "We are now 10 minutes from the hotel" he updated me, "but I missed the entrance." "You know," he continued, "once I drove with my mother, I missed the entrance to the road. Then, I knew that I should not get into this road. Do you mind that I drop you here? It is no more than 200 meters to your hotel." I happily paid the 60 Euro, and an additional 10 Euro tip for the good witch that was with us.

While I was walking with my luggage up the hill to the hotel, a taxi stopped behind me. It was my big driver again… He said: "I forgot to give you my card. Please call me when you want to go back to the airport. It will be only 30 Euro. We can take the highway." I am still in Portugal for another 2 days. I will call him if it is not too cold.

26   The Overall Evaluation
of String Theory

It was time, Gina thought, to address Smolin's overall scientific evaluation of string theory. In his book, Lee lists the five main problems in physics, claims that no real progress has been made in three decades and regards it as a failure of string theory and, more generally, of the prevalent approaches to physics in the sixties and seventies, which he considers to be no longer adequate.

Gina Dec 10th, 2006 at 7:24 am

> Dear Lee, Thanks for your interesting comments. Let me try to move ahead with my critique.
>
> 1. Asserting that there has been no progress on the five major problems of Chapter 1 is incorrect or at least very problematic.
>
> Perhaps Lee's assertion can best be seen through Roger Penrose's words on the back cover of the book:
>
> "... so his [Smolin's] claim that string theory is responsible for the lack of real progress in fundamental physics for the past quarter century carries considerable weight."
>
> Penrose attributes an even stronger statement to Lee: that string theory is not only responsible for its own alleged failure but also for the lack of progress in other areas of physics!
>
> In my opinion the claim of "no real progress" should be rejected.

The first question is, of course, what "real progress" means.

Here is an example: In the early 80's Richard Hamilton proposed a program to solve the Poincaré conjecture. Looking back at (say) 1992, did Hamilton's program represent at that time "real progress" towards the solution of the Poincaré conjecture?

In hindsight, the answer is of course yes, Perelman's 2002 proof of the Poincaré conjecture relies on Hamilton's program.

I have chosen the year 1992 because it was a decade *before* the solution of the Poincaré conjecture, and I was told that at that time most experts were quite pessimistic. Of course, some people thought that the Poincaré conjecture may very well be false. Even if true, it was not clear whether Hamilton's program was a good avenue towards a solution, and even if it were, whether the technical problems could be overcome.

My definition for "real progress" requires that the theory in question have a substantial chance for making, or contributing to a definite progress. By this rule, I would have regarded in 1992 "Hamilton program" as representing "real progress" (regardless of the ensuing turn of events), just as string theory should be regarded today.

We previously discussed the interesting question of how to test the partial progress of a theory. Here are some possibilities, none of which is ideal.

a) To try to evaluate the likelihood that the theory will achieve its goals.

b) To compare the likelihood of the theory achieving its goals to current alternatives.

c) To evaluate the contributions of the theory, CONDITIONED on the assumption that it will fail to achieve its central goal.

Criterions a) and b) measure progress in terms of promise, they are subjective and can be influenced by hypes and by crusades á la Woit. Criterion c) is indirect and is based on a gloomy worst-case scenario. In terms of the latter, I am not aware of any current tentative theory in any of the sciences that is more successful than string theory.

Five Hard Pieces: Smolin's List of Five Great Theoretical Physics Problems:

1) The problem of quantum gravity: Combine general relativity and quantum theory into a single theory that can claim to be the complete theory of nature.

2) Resolve the problems in the foundations of quantum mechanics.

3) Determine whether or not the various particles and forces can be unified in a theory that explains them all as a manifestation of a single, fundamental entity.

4) Explain how the values of the free constants in the standard model of particle physics are chosen in nature.

5) Explain dark matter and dark energy.

Controversies

Controversies and debates in and around science — between researchers within the same discipline (for instance among string theorists), between competing theories, between competing fields, and between accepted scientific viewpoints and viewpoints rooted outside science — are common. Is there global warming and is it caused by high emissions of CO_2 by humans? Or is global warming perhaps a myth, or rather an established fact caused by changes in the sun's radiation, which has little to do with us? Is quantum physics correct? Can quantum computers, which have superior computation power that can crack the codes used for most commercial communication, be built? Are the teachings regarding free-market economy scientifically based? What is rationality? What is the weight of psychology in understanding economics? What is the value of mathematical tools in the social sciences? What are the limits of artificial intelligence?

What is the origin of the Scrolls of Qumran (the Dead Sea scrolls)? Were they written by a sectarian group living in a village close to the caves where they were discovered, as the dominant theory asserts? Magen Broshi, a senior Jerusalem-based archeologist who subscribes to the central theory regarding the scrolls, said once in a public lecture: "There are at least twelve theories regarding the origins of the Qumran scrolls. These theories are conflicting, which means that eleven of them are wrong; let me tell you which is the correct one." A voice came from the audience: "At least eleven!", which was followed by an immediate reply from another audience member: "You must be a logician!"

Is there a hidden code in the Bible that can reveal details about the present and can it be used to predict the future? Are we witnessing the "end of science"? The "end of history? The "end of civilization"? A major clash between civilizations? Is the theory of evolution correct? And is "intelligent design" a serious scientific alternative? Are the claims that the HIV virus is not the cause of AIDS of any merit? Do highly diluted homeopathic medications have any real effect? And is the "Mozart effect," the claim that — unlike any other similar activity — listening to Mozart's music before an exam will greatly contribute to its successful outcome, correct?

27 High Risk Endeavors for the Young

OK, thought Gina, we can move to areas in Lee's case which do not require technical understanding of physics and mathematics. Lee advocated in his book that scientists should be encouraged to attack fundamental problems early in their career. Gina disagreed.

167 — Gina Dec 5th, 2006 at 5:24 am

12. Contrary to Lee's suggestion, young scientists should not be encouraged to attack fundamental physics problems head on.

Here are three reasons. The first is the main one.

1) They will fail

Young scientists should learn the taste of failure but they should also learn the taste of success. And they have to be able to identify success: real and definite success. This you do not learn by attacking fundamental problems head-on or by engaging in very high risk projects.

2) Thinking "big" is the natural tendency

Everybody would prefer to find a substitute to quantum mechanics rather than compute a complicated integral. Going "big" is a natural tendency of scientists — especially inexperienced ones. In science, as in surfing, or skiing, or playing the piano, you first have to learn to suppress the natural tendencies in order to do things right. (And then you can try to express them again.)

Young scientists should experience the great difficulty of getting a small problem right and the great satisfaction and value of craftsmanship.

3) The fractal nature of science

While the danger of looking too much at the big picture is that you will see nothing, in science it sometimes happens that you look at your small little problem and discover something which affects the whole picture. A film located in the wrong place at the wrong time, which gets ruined, can turn physics on its head. Innocent experiments in iterating simple functions on a hand calculator may have a tremendous effect. Young (and old) scientists should be prepared that when such an opportunity occurs — they should take it (and they should also be prepared for an unpleasant crash).

"On the other hand," remarked Gina's friend, Oded, "the good scientist will know about the big picture and will keep it in mind while working on her little problem. Moreover, you must seriously think about the big picture from time to time in order to internalize it."

Gina was very happy with the many interesting topics involved in the discussion and wanted to share her excitement with others.

168 — Gina Dec 6th, 2006 at 3:27 pm

Let me interrupt the flow of elaborations on my points (5 down, 11 to go,) with a few remarks. The first is obvious. Lee's book contains many nice, interesting and even moving parts and I am specifically trying to attack some of his arguments. Of course, the physics descriptions are very interesting. Several far-reaching ideas about physics are described and, while not remotely in a position to judge them, I am in a position to appreciate them and be impressed.

Lee also presents a bold and far-reaching view of how science should be practiced which, based on Kuhn's philosophy of science, can be regarded as a practical road-map. (Kuhn gave a prominent role to scientific revolutions.) I will try to argue with this approach but at the same time I find it a beautiful line of thought.

Look what we have here beside physics and mathematics: The issues of pluralism, measures for progress, optimism, and risk-taking, especially in science; philosophy of science — scientific revolutions and scientific conservatism; incentives for progress in science, sociology of science — scientific communities, power and influence in academia. Great stuff!

Gina forgot to mention skepticism, arrogance, tolerance, beauty in science, debates and controversies, the role of analogies, and quite a few more issues. She was truly excited by the many interesting facets of the story. Gina remembered that when she was a young child, her father told her that "everything is interesting! Every subject, no matter how marginal and unimportant or boring it may look to you, Gina, when you study it in depth you will find it fascinating! Every single subject, Gina!" And Gina says: "While being a skeptical person, I took this advice without questioning it; maybe because it was from my father."

This time, Gina quite enjoyed Lee's response.

170 — Lee Smolin Dec 7th, 2006 at 6:33 am

Dear Gina, with respect to "young scientists should not be encouraged to attack fundamental physics problems head on", I agree with the dangers you point out. And I absolutely agree that the training of scientists must include problem solving skills and experience in solving doable problems. But, in spite of everything we do to try to steer young scientists to do incremental work by which they may advance their career, there remain a few who insist on breaking away and inventing and developing original ideas to attack fundamental problems head on. My question is, what do we do about them? In my experience, two things are true:

1) Such people often have a hard time making progress in their careers,

2) In the long run, they contribute a great deal to the progress of science.

To mention some examples, they invent string theory, or the idea of quantum computation, or spin foam models, they find the correct physical interpretation of general relativity or invent new approaches to the interpretation of quantum theory (the many worlds interpretation was proposed in a PhD thesis by someone who never got an academic position in spite of the theory's great influence.)

> Still, given that they are not many in number, and are also easy to iden-
> tify, I think we can do much better for them, and in doing so speed the
> progress of science.

In his book Lee refers to scientists who study fundamental big problems from new angles as "seers", contrasting them with "normal" scientists, who are referred to as "craftspeople". Gina suggested calling the former group "dreamers", and remarked that while we already have many dreams and many dreamers the difficulty is to find some "spoilers" who will look very carefully at these wild ideas, study them, and explain to those who have them why they cannot work. Where can one find these "spoilers"? she wondered.

At a later stage Lee changed the terminology he used and referred to "seers" as "valley crossers" and to "craftspeople" as "hill climbers." Gina liked these new names much better. She always imagined scientists of all kinds as some sort of mountain climbers...

Röntgen and Feigenbaum

"A film located in the wrong place at the wrong time, which gets ruined, can turn the whole of physics on its head," wrote Gina, and referred to the accidental discovery of Röntgen Rays by Konard Röntgen. The famous Mathematician Paul Erdős liked to tell the following story: A few years before Röntgen's famous experiment, another prominent scientist named William Crookes observed that leaving a photosensitive film near the cathode-ray-tube causes damage to the film: it becomes exposed. He concluded that "Nobody should leave films near the cathode-ray-tube." Röntgen observed the same phenomenon a few years later and realized the significance of his discovery, that indeed changed physics and had a great impact on medicine.

Gina's reference to "Innocent experiments in iterating simple functions on a hand calculator may have immense effect", is a somewhat inaccurate reference to the work of Mitchel Feigenbaum on chaos. Computations with a small HP-65 computer led Feigenbaum to discover in 1975 universal properties of chaotic dynamics and fractals.

28 Scientists and Mountain Climbers

"… with a normal job, there is even a chance you will pay attention
to your wife and children when they talk to you, rather than day
dream about physics"

attributed to Lee Smolin, in Gina's story

Gina thought about the young scientists and compared them to mountain climbers, with whom they share the sense of beauty evoked by undiscovered territories, by the loneliness, the prolonged difficulties, the need for stubbornness and by the need for flexibility. There are dangers in both professions; crevasses, for instance, are a prominent hazard for mountain climbers. These, according to Wikipedia, are the slits or deep chasms formed in the substance of a glacier as it passes over an uneven bed. They may be open or hidden. In the lower part of a glacier the crevasses are open. Above the snow-line they are frequently hidden by arched-over accumulations of winter snow. The detection of hidden crevasses requires care and experience. Scientist crevasses are not deadly but they exist, Gina thought, as does the danger of getting lost and of losing one's skills. Gina's objection to encouraging young scientists to take high risk endeavors was part of her view that they should not be encouraged *at all*. Here is how Gina fantasized a meeting between Lee Smolin and a young PhD candidate called Jeremy would transpire.

Professor Lee Smolin: Please sit down. (Looks at Jeremy's file) I understand you graduated last year from Harvard University... straight A's, ... hmm junior thesis on the paper: "On the relationship between quantum and thermal fluctuation," What brings you here, Jeremy?

Jeremy: I would like to write a PhD thesis under your supervision.

LS (looks at the file, this is the best file he has seen in years...): I see. Let me tell you something right away Jeremy. Success as an undergraduate does not always mean success in research, but looking at your file I am willing to take the risk and say that I think you will be able to write a good PhD thesis.

J: Thank you.

LS: You must know that an academic job afterwards is not guaranteed. It depends on your success in research but *also* on many other factors.

J: I am aware of that fact Professor Smolin.

LS: (after a long pause). Young man, let me ask you a question. You did brilliantly at Harvard and you can *really* succeed in life. You can get out of here in one year with a Master's degree and then go and become a successful man. If you want to be rich, you can go into business or high-tech, and if you want to build things you can go into engineering, and if you want to help people you can go to med school. And wherever you go, you will be surrounded by, and working with *real people*, not only with formulas and computers. And, you know what, Jeremy, with a normal job, there is even a chance you will pay attention to your wife and children when they talk to you, rather than day dream about physics...

J: But I want ...

LS (interrupts): Look, don't give me an answer right away. Think about it. Take two or three weeks to think about it, Jeremy, and I will be happy to continue our conversation then.

Lee of course knows that Jeremy will most likely come back, and will complete an excellent PhD thesis. And then at those times when he is cold, wet and lonely on these mountains he will remember that going to graduate school was not something he was dragged into, but was truly his choice.

29 How to Treat Fantastic Claims by Great People

Anatoli Boukreev, Gina remembered, was the greatest mountain climber in the world. An article in a mountain climbing journal challenged his greatness and attributed his success to his practice of climbing up the easy side of the mountains when the weather was good. Boukreev ascended the next mountain from the difficult side. In the winter of 1997, Boukreev was attempting to climb the south face of <u>Annapurna I</u> (8,078 m) and was hit by bad weather. He did not survive.

The next topic of discussion was about the reaction of the scientific community to bold and speculative theories and scientific suggestions. In his book, Lee Smolin describes how revolutionary scientific claims are dismissed by the community even when they are proposed by great scientists. As an example, Lee describes the dismissing approach of most physicists towards the recent ideas of Gerard 't Hooft, a Nobel Laureate and one of the greatest physicists of our time.

288 — Gina Dec 16th, 2006 at 11:38 pm

Let me continue with the elaborated critique of Lee's book.

13. The negatively depicted reactions to sporadic bold theories and ideas are, in fact, very reasonable. Scientists' current attitude is essentially correct.

The attitude of the high-energy physics community towards very revolutionary scientific ideas and theories, even when these are proposed by great scientists, is portrayed negatively in Lee's book. I beg to disagree:

The basic principle in scientific and academic activity (unlike other areas) is that everybody, no matter how famous, has the burden of convincing the community from scratch about the merits of a new theory or a new idea that she or he presents. Everybody, Smolin, 't Hooft, Witten, Penrose, Chomsky, … Everybody.

Amazing Possibilities

Understanding our fundamental limitations is among the most important contributions of science and of mathematics. At the same time, various fundamental limitations stated by many great minds turned out to be wrong, sometimes rather quickly.

There are quite a few cases where things that were considered to be impossible turned out to be possible. Immanuel Kant claimed: "No finite Reason can hope to understand the production of even a blade of grass by mere mechanical causes." This quote is from the *Critique of Judgment* (1790). Elsewhere Kant wrote: "It is absurd to hope that another Newton will arise in the future who shall make comprehensible by us the production of a blade of grass according to natural laws which no design has ordered."

Auguste Comte claimed: "Of all objects, the planets are those which appear to us under the least varied aspect. We see how we may determine their forms, their distances, their bulk, and their motions, but we can never know anything of their chemical or mineralogical structure; and, much less, that of organized beings living on their surface ..." (*The Positive Philosophy*, Book II, Chapter 1 (1842)).

Spectroscopy was developed by Gustav Kirchhoff in the 1840s, and the first spectroscopic analysis of the Sun appeared about ten years later, less than 20 years after Comte's statement.

A slightly different example relates to the philosopher Wittgenstein. As usual he is more cryptic. He gives an example of a person making claims that seem crazy to us. Writing before expeditions to the Moon became a reality, Wittgenstein speaks of a person who claims that "men sometimes visit the Moon." The following excerpt is taken from "On Certainty."

"What we believe depends upon what we learn. We all believe that it isn't possible to get to the Moon; but there might be people who believe that it is possible and that it sometimes happens. We say: these people do not know a lot that we know. And, let them be never so sure of their belief, they are wrong and we know it."

There are famous quotations of predictions made by Lord Kelvin. Kelvin did not believe that heavier-than-air flying machines were possible and he regarded X-rays as a hoax. Kelvin's ingenuity was manifested even in cases where his overall predictions were wrong. He gave a lecture on the state of physics at the turn of the twentieth century, and — not unlike Hilbert's famous lectures in mathematics — claimed that physics was nearly complete and all problems would soon be settled. He mentioned, however, "two clouds on the horizon", the unexpected behavior of ether in the Michelson–Morley experiment and the problem of the spectrum of the black body radiation. His genius as a physicist was manifested by the fact that of all the scores of open problems in physics present at the time (as there always are), he pinpointed the two problems that subsequently led to revolutions: the ether problem led to relativity, and black body radiation to quantum theory.

[Editor's note: the items and citations of Kant, Comte, Kelvin, and Wittgenstein were contributed by Itamar Pitowsky, a philosophy professor at the Hebrew University of Jerusalem and my former classmate there. Itamar also drew my attention, in the summer of 2006, to the weblog discussions regarding string theory, and especially to Peter Woit's weblog.]

30   Sociology of Universities

Old people are simply young people that have turned old along the years.

Gina, Dec 5, 2006.

Sociology was a big issue in Lee's book and Gina tried to understand his point of view.

166 — Gina Dec 5th, 2006 at 5:15 am

Let us move to general matters of universities and academia.

15. Some natural, often welcome social processes in academia (regarding academic judgment and decision making) are depicted negatively. This is a mistake.

Here, I refer to Chapter 19 in Lee's book that deals with the sociology of universities and academic life.

Let me quote a few sentences and add some brief comments:

1. "The idea of changing the way science is done in universities will no doubt appeal to some, while horrifying others. But it's probably in no danger of happening. To understand why, we need to inspect the dark underbelly of academic life. Because as the sociologists tell us, it is not just about wisdom, it is about power: who has it and how it is used."

I find it impossible, even after reading the whole chapter, to understand what is said in these sentences which open the chapter and which draw a dark and gloomy picture of universities, of the way science is done in universities, and their inability to change.

2. "In the academic world, with few exceptions, the people who evaluate you are older than you and more powerful."

This may seem bad at first, but it is actually not the case. Overall, I think that it is very reasonable that older people should have more influence.

I do not see anything wrong with old people. My perception is that older people are simply young people that have turned older along the years.

Mathematician Daniel J. Kleitman in 1967 (left) and around 2000 (right).

Curiously, part of the reason for this observation by Lee is not sociological at all, but purely statistical. If you are young, the people who evaluate you tend to be older, and if you are old the people who evaluate you tend to be younger.

3. "... a process of enforced consensus in which older scientists ensure that younger scientists follow their directions"

When there is a need to make collective decisions, the notions of power and influence automatically emerge. It is overall positive rather than negative that the amount of influence is positively correlated with age and with prominence. I am not aware of any mechanism which will lead to making the optimal decisions or which can eliminate the possibility of abuse of power. If Lee has ideas in this direction it would be great to hear them.

4. "A professor will shamelessly write letters slanted towards his or her own students"

Yes, many people do various things (usually fully legitimate but sometimes a little dubious,) to promote their points of view and many are biased towards their own opinions, style, direction of research, and even students. This is a positive rather than a negative fact of life.

Reading a book is always an emotional experience and not just an intellectual one. Lee's discussion in Chapter 19 conveys an obvious mood of disappointment, and, to some extent, despair of universities, of the way they do science, and of their ability, or lack thereof, to change.

In his reply, Lee explained his position starting with his ideas on mechanisms for collective decisions.

To Gina,

I made a number of suggestions in my Physics Today article as well as in the book. But these are standard issues that concern business, as well as academia. My impression is that we in academic science are behind the curve in several of them. For example, in investment banking, when considering the promotion of X, they tell me they do a "360", which means they seek the views of all who work with her, at both less and more senior levels. In law firms, banks, newspapers, and other businesses people who hire and manage other people are required to go through training aimed at combating these problems. In investment banks and successfully managed endowments they deliberately choose fund managers with a variety of approaches and views of the market, in order to guarantee a diversified portfolio.

Gina was especially skeptical about taking investment banks as role models. [Editor's note 2016: Indeed the 2008 economic crisis and the fall of many investment banks may support Gina's skepticism in this case.]

Lee continued to describe different forms of collective decisions in universities and concluded with his belief that matters can be changed.

My point is that we should never regard the "sociology" of a field as something we just have to live with; it is the result of concrete choices about the practices that guide hiring and promotion. So if people perceive problems with the sociology of their field, rather than complaining about it over lunch, as one commonly hears, one should work within one's department or university to change the practices. One might even take a comparative approach and consider the different practices around the world as experiments, whose results are to be evaluated.

Thanks,

Lee

Power and Collective Decisions

"It is not about wisdom but about power," said Lee. What is the wisdom regarding power when it comes to collective decisions like those in university committees or in general elections? Can we measure power? Can we plan a social mechanism that will be based on evenly spreading power and will lead to satisfying collective decisions?

In the 1950s and 60s, two important mathematically-based notions of power were introduced. The Banzhaf power index for a member in a committee or a voter in an election was based on the probability that this member has the decisive vote after the other members have cast their votes. The Banzhaf power index is related to the "Penrose method," proposed by Lionel Penrose (Roger Penrose's father), for determining the number of delegates for every county in a world-parliament, with the objective of giving each county power proportional to its population. The power index proposed by Shapley and Shubik derives from a more general concept in game theory and is based on a set of axioms. It can also be based on the probabilities of members or voters being decisive.

Even in a context where these indices of power do not seem appropriate, the principle of relating power to decisiveness can still be useful.

The idea of using the language of mathematics to describe and study issues of "political science" goes back to the French mathematician, philosopher, and early supporter of democracy and feminism, Marie Jean Antoine Nicolas de Caritat, Marquis de Condorcet (1743–1794). Condorcet is famous for his "paradox" asserting that in an election with three candidates — Alice, Bob, and Carol — it is possible for a majority of voters to prefer Alice to Bob, for another majority of voters to prefer Bob to Carol, and for yet another majority of voters to prefer Carol to Alice.

In 1951 Kenneth Arrow found a far-reaching extension of Condorcet's paradox, and showed that under some simple conditions, the only voting method where this paradox is avoided is dictatorship: the outcome is always determined by the same individual. The interpretations of these particular results, as well as of issues regarding mathematical modeling in the social sciences, are subject to intense controversy.

31　The String Theory Community

The next item, Lee's description of the string theory community, was perhaps the most delicate one. Gina gave considerable thought as to how to express her criticism in the correct manner.

206 — Gina Dec 14th, 2006 at 12:33 am

> 9. Many claims, stories and quotes against the string theory community are not really damaging to the ST community and bringing them weakens the quality of Lee's argument.
>
> Let's look at some specific claims (mainly from Chapter 16 of Smolin's book) against the string theory community:
>
> Lee talks a lot about "group thinking" and about the need "to fight" group thinking. It is not clear what he means precisely. Research based on large collective pools of ideas and results is a very positive phenomenon which characterizes string theory and other prominent research areas. When exactly does this blessed and crucial practice become the negative "group thinking"?
>
> I especially disliked the analogy drawn by Lee between "group think" in string theory and "group think" that brought about NASA's failure to prevent the Challenger disaster, and "group think" that led to the current war in Iraq.
>
> There is also a related claim that string theory "works by fashions," so a large amount of efforts is geared towards directions which look promising. Again, I see nothing wrong with it. It reflects string theory's methodology to try to send long and thin in-depth sensors and

get some picture of what the emergent big theory looks like, while neglecting or delaying other directions due to technical difficulties.

The repeated claim of "arrogance" does not make much sense either, and it is not really supported by quoting a specific physicist (among thousands of physicists) who made a similar claim.

Overall, the feeling I got reading the book was that Lee's description of the string theory community is unfair and incorrect.

32　The Finiteness-of-String-Theory War

The discussion on Mandelstam's work ignited an extensive debate regarding the results achieved by Mandelstam himself as well as by Atick, Moore and Sen (AMS), and by d 'Hoker and Phong, who all assert the "finiteness of string theory". Gina did not consider the weblog environment as suitable for a technical discussion on this matter, and she was rather interested to know (but did not get an answer) whether Mandelstam was related to the famous Russian poet Osip Mandelstam. The debate spanned over 100 comments for several weeks with a lot of heated hair splitting. The debate also spilled over to other blogs. Lubos Motl gave an interesting account of the finiteness issue on his blog, and offered Jacques Distler a "medal for patience". Lubos' post was characteristic of his bold approach towards string theory and mathematics. His overall view was that, with the progress and triumph of string theory, the mathematical difficulties in mathematical physics (and perhaps in all mathematics) would be resolved as well. Gina used the opportunity for a quick interview with Jacques.

245 — Gina Dec 14th, 2006 at 11:13 pm and **253 — Jacques Distler** Dec 15th, 2006 at 7:37 am

> [**Gina**] Dear Jacques, congratulations on the new medal for patience. Let me try and stretch your patience a little and ask a few questions:
>
> 1. Is it correct to say that neither the work of AMS, nor even that of d 'Hoker and Phong can be regarded as solid proof in the mathematical sense of the word.
>
> [**Jacques:**] I don't think physicists strive for (nor should they) a level of rigour characteristic of Mathematicians. Nevertheless, with a finite

(but sometimes very large) amount of effort, it is usually the case that a "Physicist's Proof" can be turned into one that would satisfy a Mathematician.

[**Gina**] 2. According to your explanation, the new proof is "constructive", while AMS proof is not. Can you explain in non-technical terms what the difference is, and why it is of importance to find constructive proofs?

[**Jaques:**] A non-constructive proof says: even though we cannot write down the answer explicitly, we *can* prove that it has properties A, B and C. And this is sufficient to show that desired outcomes X, Y and Z are achieved.

A constructive proof says: Look! We can write down the answer explicitly and you can see, by inspecting it, that it has the desired properties.

Constructive proofs are always preferable in Mathematics, but even *more so* in Physics, where we would like to use the answer to calculate stuff.

[**Gina:**] 4. Smolin's concerns on this issue were that string theorists were too optimistic to interpret partial evidence (Mandelstam's) as definite. Do you agree with Smolin at least on this point?

[**Jacques:**] I *really* don't understand Smolin's account. Mandelstam's announcements generated a lot of excitement, but, when he never wrote a paper about it, everyone figured that he had found a flaw in his method.

[**Gina:**] 5. What is your take concerning this "divorce" of physics from mathematics. (I learned this term from a paper by David Corfield posted on the n-category cafe.) Is it something physicists should worry about at all?

[**Jacques:**] As to the larger point, there are many things in physics that we "know" are true, because we have amassed a lot of computational evidence to support them, but which have never been formally proven.

It was especially interesting to learn from Jacques about the importance of proofs in physics for providing effective tools of computation.

Gina was overall quite skeptical when she heard scientists talking about "solutions" and "proofs." Ever since she heard the nice story, that we tell in the next chapter, about Jeffrey and Miles, she asked herself: Are solutions really solutions?

Debates

Debates are fascinating human activities that are a mixture of logic, strategy, and show. Not everybody shares this fascination. The German author Emil Ludwig considered debates to be the death of conversation. Jonathan Swift regarded debates as the worst sort of conversation, and debates portrayed in books, as the worst sort of reading. Public debates pose various interesting dilemmas. A debate between two positions gives an impression of symmetry, and engaging in a debate against an obscure or illegitimate position gives it some legitimacy and emphasis. On the other hand, ignoring obscure or illegitimate positions may also pave the way to giving them public legitimacy or to making them mainstream. A common form of debate is one in which an uninformed decision-maker extracts information from two (or more) informed debaters who hold contradictory positions on a certain issue. Weblog debates are especially interesting, as they allow an unusual amount of interaction between the debaters and the uninformed audience. Common debate practices include ample repetitions, not giving up on seemingly small issues, never admitting a mistake, and trying to undermine the professionalism and integrity of an opponent and not just his logic. We can ask ourselves if these debating practices are rational and optimal in terms of influencing the audience. They probably are.

33 A Mathematician's Solution to the Proliferation Problem

"Modeling, modeling, modeling. If you want one important notion in this business it is mathematical modeling. The art of modeling, the science of modeling."

Miles, around 2002.

Miles does not look well, thought Jeffrey. Miles was a young mathematician at Oxford. He was working on something called K-theory and Jeffrey could not figure out what it was about. Miles was a good mathematician, at some point he was the youngest reader in the whole kingdom. Jeffrey was used to seeing Miles troubled at times by a math problem and it was difficult to talk to him on such occasions. But this time Miles looked even worse than usual.

"What's on your mind, Miles?" asked Jeffrey.

"Well Jeffrey," answered Miles, "I am deeply troubled by the problem of proliferation."

"Proliferation?" asked Jeffrey "not that I intend to understand, but tell me: What does proliferation *mean* in K-theory?"

"No, no" said Miles "I am talking about the problem of proliferation of nuclear weapons."

"I see," said Jeffrey.

Bad news, thought Jeffrey. Proliferation is a troubling problem indeed, but he had been around long enough to realize that getting really upset over it was a bad idea. And from Miles' look and his rather incoherent thoughts on the matter, it all looked like a one way street.

Fourteen months later Jeffrey met Miles again. To his surprise Miles looked completely fine and even quite cheerful. "What's up, Miles?" he asked. Miles got right to the point. "Well, you remember that I was troubled by the problem of proliferation, I spent much of my time in the last months studying this problem. This was a difficult project that led me to many interesting areas."

"But, Miles" asked Jeffrey, "I never thought about this problem as a problem in mathematics. It is a problem about politics, isn't it? Can you come up with a formula in mathematics which has something to do with proliferation?"

"Not a formula", said Miles, "but a model."

"Modeling, modeling, modeling" Miles continued "if you want one important notion in this business it is mathematical modeling. The art of modeling, the science of modeling."

Then he paused,

"Jeffrey", asked Miles "did I mention '**modeling**'?"

Jeffrey was used to Miles' strange sense of humor. "Yes, Miles, you did mention modeling."

"Building the relevant mathematical models was the most important part in my project," said Miles.

"But Miles", said Jeffrey, "this whole issue involves a lot of uncertainties. You cannot know how things will turn out."

"You are absolutely right," said Miles. "Probability, probability probability," he continued, "if you want one important notion in this business it is most probably the notion of probability. The art of probability, the science of probability."

Then he paused,

"Jeffrey", asked Miles "did I mention '**probability**'?"

"Yes, Miles, you did mention probability," said Jeffrey

"A major part of my project was in the realm of probability and statistics. Without probability I would not have the language to talk about the problem at all."

"But Miles," said Jeffrey "this issue is related to *people*, and how people behave. How can you translate all this into mathematics?"

"Rationality, rationality, rationality," said Miles "if you want one important notion in this business it is rationality. The art of rationality, the science of rationality."

Then he paused,

"Jeffrey", asked Miles "did I mention '**rationality**'?"

"Yes, Miles," said Jeffrey "you did mention rationality."

"A careful analysis taking into account the multiple approaches to this notion of rationality, irrationality and bounded rationality was a major part of my project, Jeffrey."

"But Miles" said Jeffrey, "When you analyze how one person or country behaves, you have to take into account not only what this agent prefers the most, but also how his action affects the behavior of other agents. This looks very complicated."

"What you are talking about, Jeffrey, is called strategy," said Miles, "Strategy, strategy, strategy," Miles continued "if you want one important notion in this business it is mathematical strategy. The art of strategy, the science of strategy."

Then he paused,

"Jeffrey", asked Miles "did I mention '**strategy**'?"

"Yes, Miles, you did mention strategy," said Jeffrey.

"Game-theoretic strategic models played an important role in my work, Jeffrey."

"But then Miles, when you model the behavior of the different agents and take into account strategic behavior, what comes next? How do you actually solve the problem?" Jeffrey asked.

"Optimization, optimization, optimization," said Miles, "if you want one important notion in this business it is mathematical optimization. The art of optimization, the science of optimization."

Then he paused,

"Jeffrey", asked Miles "did I mention '**optimization**'?"

"Yes, Miles, you did mention optimization," said Jeffrey

"Once modeled, all these issues of rationality and strategy translated into mathematical optimization problems. This was a major part in my project, Jeffrey."

"But tell me Miles," said Jeffrey, "what is the meaning of all these models and mathematics?"

"Interpretation, interpretation, interpretation," said Miles "if you want one important notion in this business it is mathematical interpretation. The art of interpretation, the science of interpretation."

Then he paused,

"Jeffrey", asked Miles "did I mention '**interpretation**'?"

"Yes, Miles, you did mention interpretation," said Jeffrey

"Jeffrey," said Miles "I built probabilistic models and took into consideration matters of strategy and rationality, I set the relevant optimization problems, and, being cautious of misinterpretation and over-interpretation and the effect of noise, I can say that I feel quite comfortable now ..."

"Do you have a solution?" interrupted Jeffrey. This all sounds crazy, thought Jeffrey, but many people will be interested in a solution.

"Let me just say," said Miles "that I feel pretty good about my overall approach and the emerging insight, yes, you can call it a solution."

"But what *is* the solution?" asked Jeffrey.

"The emerging solution is simple" said Miles "and I can almost say it is elegant," he continued. "Rather than bore you with mathematical formulation and details, let me try to describe my solution in non-technical informal terms. We are all going to perish."

34   Shocking Revelations

Clifford, I acknowledge that this was something I should not have done and apologize for it.

— Peter Woit, Dec 16th, 2006 at 9:25 am

With more and more comments and very heated debates, a discussion between Clifford and the blogger named Christine Dantas caught Gina's attention. Christine Dantas was herself a well known blogger and had a blog devoted to quantum gravity and other topics in physics called "Christine's background independence". Reading the exchange between Christine and Clifford, Gina could not escape the thought that Clifford did not read Smolin's and Woit's books, so she decided to ask him about it.

262 — Gina Dec 15th, 2006 at 2:35 pm

Dear Clifford,

Did you read the books by Peter Woit and Lee Smolin? Sure, there are various aspects and details of these books that can be criticized, but what you wrote is an oversimplification of these books and, even more than that, of the many interesting issues involved in this discussion.

263 — Clifford Dec 15th, 2006 at 7:03 pm

Hi Gina,

Please read what I've been saying on this issue–in great detail–in discussions with Smolin and Woit for the last year and a half. I know their positions very well, and, overall, I don't agree with them.

… I have not read their books, but it is well known that those views are also represented in their books. I am free to make my own independent objections to their publicly expressed views, and I have made them. I did not write any book reviews, which would be wrong, since I have not read their books.

(By the way, contrast that statement with the fact that Woit made up out of whole cloth the public statement/accusation that I negatively reviewed his book for CUP (Cambridge University Press). If he makes up stuff like that at random, how can a non-expert trust what he has to say about the host of other things he says, about research in the field of string theory? You don't make up things to cast a bad light on someone's professional reputation (he also accused me of somehow influencing KC Cole to write her LA Times review!) and put them on a blog.)

Best,

-cvj

Gina suspected all along that Peter Woit got it wrong regarding the identity of the Cambridge University Press referee, but she did not expect that the factual matters would be unfolded and certainly did not expect to play a role in such a discovery.

271 — Peter Woit Dec 16th, 2006 at 6:53 am

Clifford,

Thanks for making it clear that you were not the CUP referee, had nothing to do with the KC Cole "review", as well as acknowledging that you haven't even read my book or Lee Smolin's. My apology for having made erroneous suggestions in my posting, and for misinterpreting your later comments about it. I will update that posting to include this apology.

So, as it turned out, Clifford did not read Woit's and Smolin's books and Woit wrongly accused Clifford of being the hostile referee of his book for Cambridge University Press. Some factual matters were cleared up after all, albeit not being directly related to the blueprint of our universe.

Gina felt good about her participation, despite a comment by a blogger called "anon on the Hudson", who accused her of blog-trolling. "Am I a blog-troller?" thought Gina.

During the last two weeks of 2006 Gina continued to present her 16 points to Lee. Her last comment addressed Lee's ideas about philosophy of science. Gina did not like Lee's call for an academic environment that would encourage scientific revolutions. In fact she regarded "progress" rather than "revolution" as much more characteristic of what science is and quoted philosopher Sidney Morgenbesser who, making fun of Kuhn's theory, wrote: "Brother, can you spare a paradigm?"

By New Year's eve of 2007 Gina was done and she wished everybody a happy 2007. She was quite relieved.

Who is Gina II

by Anat Amirav

It was one of those typically hot Israeli end-of-August days; a scorching summer morning, where you have to convince yourself that the cool breezes of autumn are just around the corner.

To escape the unbearable humidity of the coastal area, I decided to treat myself to a visit to Jerusalem. My cause for excitement was twofold: not only would I enjoy the sites and atmosphere of this beautiful city, but this impromptu visit would also be the perfect opportunity to meet with dear Prof. Gil Kalai, the mastermind behind these very pages, a man who boldly followed Gina's adventures through cyberspace, and who cleverly pieced them together into a highly entertaining book.

Little did I know, however, that this would not be just another visit to Jerusalem, or that this seemingly typical summer day would turn out to be anything **but** typical. Indeed, it was to be a memorable day of SHOCKING REVELATIONS!

Shortly after arriving at the Hebrew University's scenic Givat Ram campus, I made my way to Prof. Kalai's office. After we exchanged pleasantries and discussed our respective summers, our conversation naturally turned to Gina's Adventures. Gil had recently added the book to his blog, and he shared with me some of the most notable and amusing comments that had been posted in reaction to the book.

Laughing, Gil told me that one of the bloggers had even "accused" him of being Gina. This comment prompted me to ask if the "real" Gina had, in fact, attempted to contact Gil — it only seemed natural that she would have something to say to the man who had so blatantly "outed" her.

I'm not sure I recall Gil's precise response to my question. Whatever it was, it faded into his next — almost nonchalant — comment: "Well, the blogger *was* right. I *am* Gina".

I felt I was in a classic "who dunnit" Agatha Christie-like moment, when Hercule Poirot or Miss Marple have just exposed the identity of the ingenious culprit, whom no-one had even suspected.

Truth be told, apparently some bloggers had correctly interpreted the clues; but as for me — well, this confession could have absolutely knocked me over with a feather!

Gil is Gina, Gina is Gil!

A million thoughts and emotions ran through me: It wasn't just the idea that Gil had created Gina, that he was, in a sense, her Pygmalion — it was also the growing realization that Gina did not really exist. She was a figment of Gil's imagination!

It was a truly sad moment for me; I had grown attached to Gina, she had made me laugh; most of the time she infuriated me; she even managed to move me once or twice. There's no denying it — for better or worse, Gina had become a friend.

There was never a doubt in my mind that she was a real, flesh and blood woman.

Well, dear Prof. Kalai, this leads me to but one inevitable conclusion: if you are Gina's Pygmalion, then she is undoubtedly your Galatea — a virtual being come to life, a phantasm turned human.

I, for one, can't wait for the movie!

35 How to Debate Beauty?

After New Year's 2007, Gina's participation in the string war weblog discussions slowed down considerably. Gina took part in another round of discussions on the blog of Clifford Johnson, who remained her favorite blogger, and participated in a marathon thread[1] (500+ comments) entitled "String theory is losing the public debate" over the blog "the cosmic variance." She discussed the role of mathematical rigor with Joe, made sure that the 500th comment is kept for Mark, talked about "Fock spaces" with Hendrik, and about the "Free Lunch Theorem" with Lee, and she even made a comparative study of the Californian cities of Santa Barbara and Gilroy. In Gilroy, the "garlic capital of the world", Gina claimed, you smell and feel garlic everywhere. But in Santa Barbara with its famous physics you do not feel or smell any physics. Gina enjoyed discussing technical matters which were well over her head, and she wrote up her long-planned remark about the debate on beauty.

GILROY
THE GARLIC CAPITAL OF THE WORLD

SANTA-BARBARA
THE PHYSICS CAPITOL OF THE WORLD

---

[1] The original post and thread of comments can be found in http://www.preposterousuniverse.com/blog/2007/03/21/string-theory-is-losing-the-public-debate/

[Debating Beauty] (cosmic variance)

The issue of beauty and physics is quite prominent in this discussion. Lee Smolin warns against adopting a physics theory based on aesthetic consideration and brings Kepler's theory relating the five planets and five platonic solids (regular polytopes) as an example. Peter Woit makes (repeatedly, again and again and again) the claim that string theory is simply ugly, very ugly.

Well, beauty is a subjective matter. I remember my dear grand uncle Lena telling me: "Gina, aren't we very lucky that people see things in a subjective way? If men were objective they would have all fallen in love with my own beloved wife (her name was incidentally also Gina), who is clearly the most beautiful woman. This could have caused all sorts of complications."

I, for example, regard string theory as very beautiful. Supersymmetry which grew up along string theory is an extremely beautiful notion. (In my view, supersymmetry has a natural form of beauty while string theory has an exotic and peculiar beauty.)

But the really interesting question in my mind is how to debate beauty. Can beauty be argued and debated at all?

Here is a story about arguing beauty in court, which may be of use. It was a case where the defendant was accused of a terrible crime.

The attorney for the defendant said in his opening speech: "Look at the defendant. Look how beautiful he is and look at his blue eyes, eyes of an angel. Do you really think he is capable of committing this ugly crime?"

At first, the prosecutor thought of ignoring this remark altogether, but then the remark was repeated and similar sentiments were expressed by some witnesses. The prosecutor watched how this non-issue was becoming an issue, and was worried that the beauty claims might convince some jury member.

The dilemma was not a simple one. Trying to argue that the defendant was not beautiful might convince a few jurors but would strengthen the belief of others that having beautiful eyes is indeed an impediment to being a criminal. Trying to argue that there is no connection between the innocent angelic look and the crime may give this whole beauty business some credibility, and may cause those jurors who believe in this connection to take for granted that the defendant is indeed beautiful.

This is what the prosecutor said in his closing argument:

"Ladies and gentlemen of the jury, there are two types of beauty. There is beauty that reveals a beautiful soul and there is beauty that covers up a corrupted and distorted personality. It is very difficult to distinguish between these two types of beauty, and often our initial hunches and intuitions turn out to be wrong.

We have carefully proved during this trial that the defendant committed the crime he is accused of, and therefore you must **conclude** that to the extent to which you find him beautiful, this is beauty of the bad kind, beauty which covers a corrupt personality capable of committing terrible crimes."

Who is the most beautiful queen of cards? Opinions vary.

Epilogue: The Real Border

And this is a small piece Gina wrote about her brother Steve, and posted on Scott Aaronson's weblog "Shtetl-optimized."

Anonymous Says:
August 8th, 2006 at 4:12 pm

The Real Border

"I see holocaust in the street" says Steve. I try to reason with him. "It did not happen here," I say, and "But Steve, our own relatives immigrated well before the war," and I even ask, "What precisely is it that you see?" But Steve only says:"I see holocaust in the street and also in my cup of coffee."

My brother is schizophrenic. His illness makes him suffer. This is an intolerable suffering which has been shattering and corrupting his identity and personality for forty years now.

On the other side of the real border there are those whose suffering is unbearable and inexplicable. The wisdom, the knowledge, the understanding, and the justice is on our side.

The more I understand, the more I succeed, the happier I am, the deeper the border between me and my brother.

[Editor's Epilogue, 2007:] The future of string theory as the definitive theory of high-energy physics is unclear. Much depends on the results of the Large Hadron Collider, which are expected in several years, and on theoretical developments in and around the theory. String theory can prevail in ways we cannot anticipate at present and it can also fail in unexpected ways. This book is not about string theory. It is more about delicate boundaries between greatness and megalomania, between humility and arrogance, between fantasy and reality, between wisdom and bullshit, between people of different stature and standing, between skepticism and harassment, between sanity and its loss, and between truths and fallacies. These are delicate boundaries that we witness in academics and in science and even in blog discussions. This story offers a small salute to people's passion for understanding their logical and physical reality, as well as for understanding themselves.

Appendix: Puzzles on Trees, High Dimensions, Elections, Computation and Noise

Gil Kalai, The Einstein Institute of Mathematics, The Hebrew University of Jerusalem.

An expanded English version of a lecture given at the Israeli Academy of Sciences and the Humanities, December 2016.

In this lecture I will talk about some mathematical puzzles that have preoccupied me over the years, and I will also reveal to you some of the secrets of our trade. The first puzzle we shall discuss is about high-dimensional trees: what they are and how to count them. The second puzzle deals with high-dimensional geometric bodies, and a question by Borsuk. The third puzzle is about errors made when counting votes during elections, and the fourth puzzle raises the question: are quantum computers possible? I will conclude with a puzzle that I am currently thinking about: random RNA trees.

Puzzle 1: How Many 2D Trees are There?

Mathematicians often use words for their own needs. "Fields" in mathematics are different from fields in nature, and "groups" in mathematics have little to do with groups of people. "Tree" is yet another mathematical concept bearing little resemblance to real-life trees.

A tree is a "graph" made up of "vertices" and "edges" with two properties: it is "connected" and it has no "cycles."

In this explanation I have replaced one concept that needs to be explained with an entire sentence consisting of several concepts that still need to be explained. I will try to explain further and with the use of examples, but I would like to point

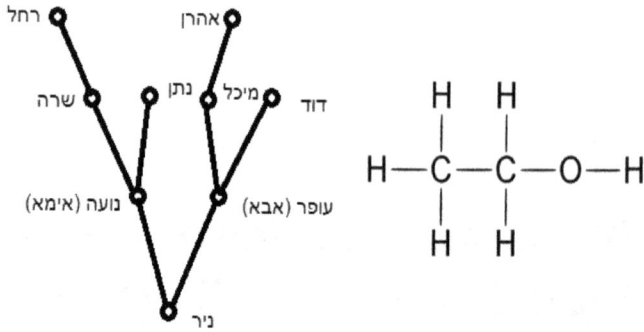

Figure 1:   Family tree (left, names written in Hebrew); the chemical formula for ethanol (right).

out that you can see my lecture as an attempt to tell an interesting story, although, due to space limitation, parts of it are bound to remain a mystery. I hope non-mathematicians will bear with me despite the fact that parts of this lecture do require some background in mathematics. Research itself can also be seen as an attempt to understand a very interesting story that, because of our limited abilities, will never be fully understood.

Anyway, Arthur Cayley, a prominent English mathematician, was able to calculate the number of all trees with n labeled vertices, and to reach the following eponymous formula:

Cayley's formula: The number of trees with n vertices is n^{n-2}.

Cayley's motivation came from chemistry, since trees are used to describe chemical molecules – the vertices are the atoms and the edges are the chemical bonds.

Trees can be seen as one-dimensional geometric structures. When I was a young doctoral student in 1980 I was interested in the possibility of defining two-dimensional "hypertrees" (and "hypertrees" of even higher dimensions) and in extending Cayley's formula to higher dimensions. In the two-dimensional trees that I tried to define, there are vertices, edges and triangular "faces." In dimension 2 we do allow the edges to create a circular structure but require that each "hole" created by the edges be filled with triangular faces. This feature is called "one-dimensional connectivity."

Even before I knew exactly how to define two-dimensional trees, I proposed the following answer for counting their number:

The number of 2D trees with n vertices is $n^{(n-2)(n-3)/2}$

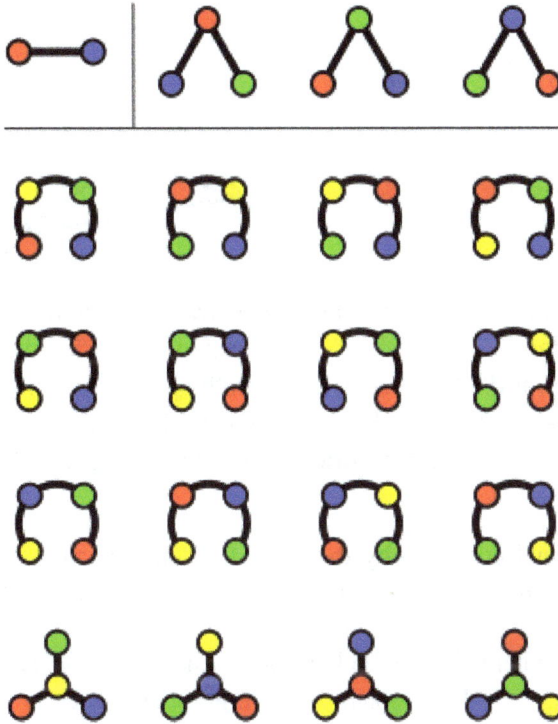

Figure 2: In the image we see all the trees with two, three and four distinguishable vertices. (The vertices are distinguished by colors.) Look at the graph on the left of the second row, whose vertices are marked yellow, green, red, and blue. If we drop an edge, say the edge between the blue vertex and the green vertex, we destroy connectivity; i.e., we get two disconnected parts. If we add an edge, say from the red vertex to the blue vertex, we get a circular structure (a cycle).

As the picture illustrates, Cayley's formula states that there is one tree with two vertices, three trees with three vertices, and 16 trees with four vertices. (Source: Wikipedia; author: Júlio Reis.)

The answer is correct for n = 3, 4, and 5, but there was a serious problem with it: it is wrong for n = 6. For n = 6, the formula gives 46,656, but I was only able to identify 46,608 "hypertrees" and another 12 dubious cases (which even if included, do not give the desired result). My way of dealing with the problem, and here I am once again revealing to you a secret of our trade, was to change the question, and adapt it to the requested answer.

What is needed in changing the question and adapting it to the requested answer opens a window into a number of central issues in modern mathematics. An important topic, related to the field of mathematics called "topology," is the topic

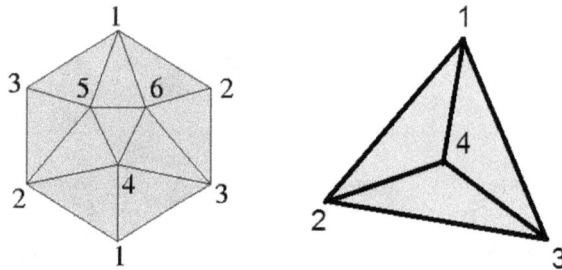

Figure 3:   On the right is a two-dimensional "hypertree" with 4 vertices, 6 edges, and 3 triangular faces. On the left is a dubious two-dimensional "hypertree" with 6 vertices, 15 edges, and 10 triangular faces. (In the picture three vertices and three edges are represented twice and one should try to imagine a fold of the geometrical shape that identifies the two appearances of those vertices and edges.)

of multidimensional connectivity that was developed in the late 19th and early 20th centuries and is associated with the mathematicians Betti and Poincaré.

Betti, an Italian mathematician who lived at the end of the 19th century, associated a number with every two-dimensional structure in order to measure one-dimensional connectivity. When Betti's number is zero the structure is "connected," in Betti's eyes!

Poincaré realized that there is a subtler parameter: a group! This group is called the homology group. When Betti's number equals 0, the homology group is a group with a finite number of elements. According to Poincaré, the body is "connected" only when this group has only one element.

Poincaré was one of the greatest mathematicians of all times, and his work also laid the mathematical foundations for Einstein's special theory of relativity. Poincaré himself was not convinced by Einstein's theory and did not accept it.

Here we meet the concept of a "group," which was a central theme in mathematical research in the nineteenth century (and ever since), as well as the concept of "homology," which was discovered and became fundamental in twentieth century mathematics.

What I found out about two-dimensional trees is that in order to get the desired formula, trees that are connected according to Betti have to be counted with weights that are the square of their homology group. The 12 dubious cases are exactly those that are "connected" according to Betti, but their homology group has two

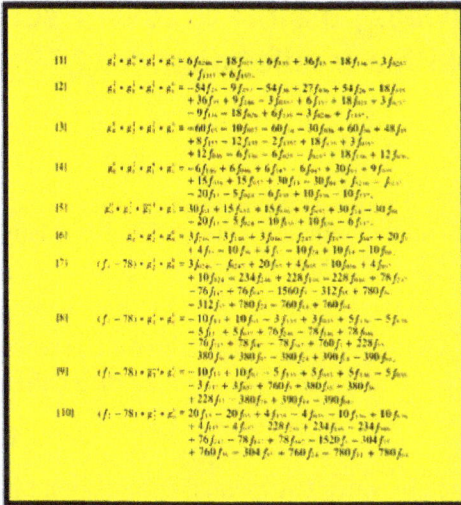

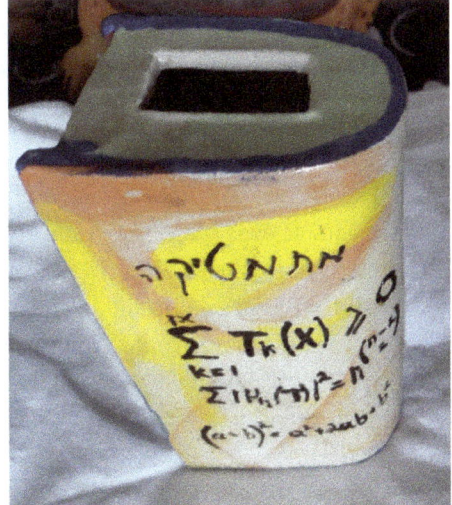

Figure 4: Bernard Benet used in his art my mathematical formula, originally intended for the study of high-dimensional objects (left), and my sister Tamar Kalai created a ceramic savings box with mathematical formulas (right). Besides the formula for 2D trees, you will probably recognize a familiar formula: $(a + b)^2 = a^2 + b^2 + 2ab$. Our father introduced this formula to me when I was six, and I was immediately fascinated by it. When I shared the story with colleagues from all over the world, it turned out that for some of them, this was a familiar experience.

elements, and therefore, they must be counted four times! The formula that I found and that made me quite happy was that for every number of vertices (and every dimension) we have to take into account the "trees" according to Betti's definition, but in those cases where there is disagreement between Betti and Poincaré, they should be counted with a weight equal to the square of the size of the homology group. I was very happy when I found a simple-looking mathematical formula that describes the result in its entirety:

$$\sum |H_{k-1}(K)|^2 = n^{\binom{n-2}{k}}$$

The concept of trees is important in many branches of mathematics, and mathematicians also deal with infinite trees and their multidimensional extensions (called "buildings"). The connection between discrete structures (like trees) and continuous structures is important in all of our puzzles, and my colleague Ilya Rips defined an important continuous concept of "real trees." And finally, with Nati Linial and Yuval Peled, we have recently come to grips with the original question of counting two-dimensional trees without weights.

Puzzle 2 (Borsuk, 1933): Is it always possible to cover a set in n dimensions by n + 1 sets with a smaller diameter?

Let us now turn to the second puzzle, where we sail off to the world of high dimensions.

Karol Borsuk was an important Polish mathematician known for his many contributions to the field of geometry. Among other things, he discovered a geometric body called "Borsuk's trumpet" that is connected to the two-dimensional trees that we discussed in the first puzzle. During the Second World War, to make a living, Borsuk invented and distributed a board game with 12-sided dice called Animal Husbandry.

The diameter of a set is the greatest distance between two points in the set, and in 1933 Borsuk asked if it is always possible to cover a set in n dimensions by n + 1 sets with a smaller diameter. Borsuk realized that in two dimensions the answer to his question is positive and follows from a result from 1906 on packing shapes in hexagons. He believed, as did many others, that the answer was positive in every dimension.

Some clarifications on high dimensions. The space we are familiar with is a three-dimensional space; what, then, is, the meaning of spaces of larger dimensions? Based on the relationship between geometry and algebra developed in the 17th century by René Descartes, this is a surprisingly easy question for a mathematician to answer. Descartes understood then – what is now obvious to us – that a point in a plane can be represented by two coordinates (two numbers), and a point in space requires three coordinates. Why stop at three dimensions? It is possible, with ten coordinates, to describe without difficulty 10-dimensional geometry, and with 100

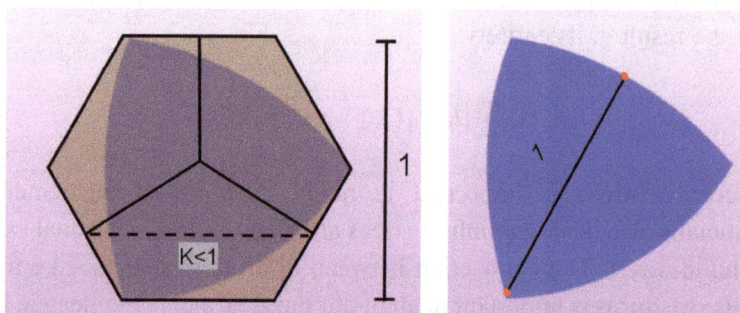

Figure 5:   On the right we see a geometrical shape of diameter 1. On the left we see how this set can be put inside a hexagon and the hexagon can be divided into three pentagons with a smaller diameter.

coordinates to describe 100-dimensional geometry. With algebra, it is also possible to define a distance between two points in high-dimensional spaces, and with algebraic formulas to describe geometric objects. I remember my colleague Yoram Lindenstrauss' first lecture at the Israeli Academy of Science and Humanities in 1985, in which he described the mysteries of spaces of infinite dimensions. I also remember many lectures about spaces with fractional dimensions (some of which are called fractals). You, the reader, might ask yourself "can the dimension be negative"? or "can the dimension be a tree?" While I do not have good answers I can assure you that these seem like questions that a mathematician would ask.

Anyway, let us return to Borsuk's question. In the early 1990s, I worked on Borsuk's question with Jeff Kahn, an American research associate. Sometimes we tried to prove that the answer was positive, and sometimes negative. Jeff was about to visit Israel when we exchanged the following amusing email.

Gil: What will we do on your visit to Israel?

Jeff: We'll solve the Borsuk problem !!

Gil: And what will we do in the second week ?!

Jeff: We'll write the article.

Surprisingly, that's exactly what happened. Jeff and his family arrived on Monday and their visit was fraught with mishaps. Only on Saturday night did we meet to discuss the question and we thought of a way to solve it. The next day we continued to discuss it at the Belgium House, our faculty club at the Hebrew University, and things gradually seemed simpler until we finally reached a solution. We found a

Dimension four	Dimension three
Hyper-cube	Cube
?	
$0 \leq x \leq 1$ $0 \leq y \leq 1$ $0 \leq z \leq 1$ $0 \leq u \leq 1$	$0 \leq x \leq 1$ $0 \leq y \leq 1$ $0 \leq z \leq 1$

Figure 6:   When describing the 3D cube with algebra, it is easy to extend the definition to four dimensions. It is not easy to visualize the resulting set in four dimensions. Try it!

high-dimensional construction (over 2,000 dimensions) that shows that the answer is negative.

For mathematicians our construction can be described in 10 words: "a tensor product of the n-dimensional unit sphere with itself."

The proof is also simple and stretches over half a page. But as short and simple as it was, it was difficult to reach. Jeff, who majored in English, aptly described the situation with a quote from Hermann Melville's *Moby Dick*:

"However contracted, that definition is the result of expanded meditation."

We used two ideas in the proof. The first idea is to move from a continuous problem to a discrete problem involving graphs. The second idea is to take the asymptotic approach and analyze the situation in high dimensions. Both ideas are closely related to a famous 20th-century mathematician, Paul Erdős.

In our next puzzles we shall talk about mathematical models and problems with real-life applications.

Puzzle No. 3: What are the methods of election that are immune to errors in the counting of votes?

In my third puzzle, mathematics encounters practical uses: the understanding and planning of electoral methods.

We all remember the sight of vote recounts in Florida in the 2000 US presidential election. Is the American electoral system, based on electoral votes, inherently more susceptible to mistakes than the majority system? And what is the most stable method? Together with Itai Benjamini and Oded Schramm we investigated these and similar problems. This was the question that we asked: given that there are two candidates and each voter chooses at random and with equal probability (independently) between them, what is the stability of the outcome when in the vote-counting process one percent of the votes is counted incorrectly? (In mathematics, these errors are called "noise.") We defined a measure of noise sensitivity of electoral methods and found that weighted majority methods are immune to noise, namely, when the probability of error is small, the chances of errors will diminish. We also showed that every stable-to-noise method is "close" (in some mathematical sense) to a weighted majority method. In later work, O'Donnell, Oleszkiewicz, and Mossel showed that the majority system is most stable to noise among all non-dictatorial methods.

Our work was published in 1999, a year before the question appeared in the headlines in the US presidential election, and it did not even deal with the subject of elections. And here one can witness another secret of our trade. Mathematicians recycle their models, and the same model can be used for very different purposes. We were interested in understanding the problem of planar percolation, a mathematical model derived from statistical physics illustrated in Figure 8. In our article we showed that if we adopt an electoral system based on the model of

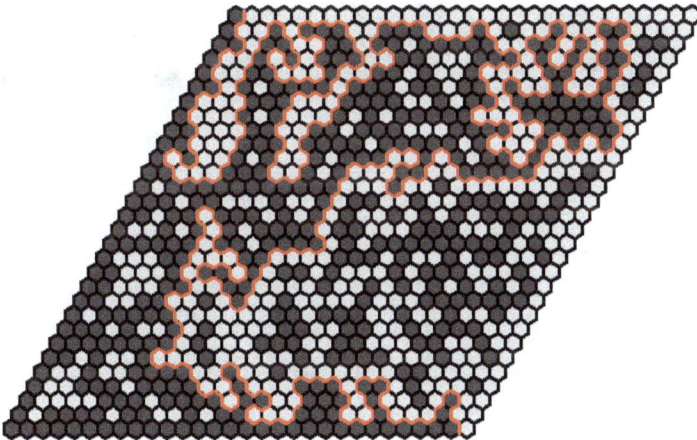

Figure 8: The model of percolation in the plane: each hexagon is grey with probability half and white with probability half. The (red) border line between the grey and white areas has dimension 4/3.

percolation, this method will be very sensitive to noise. This insight is of no use at all in planning good electoral methods, but it makes it possible to understand interesting phenomena in the study of percolation.

After the US presidential election in 2000 we tried to understand the relevance of our model and the concepts of stability and noise in real-life elections: is the measure for noise stability that we proposed relevant, even though the basic assumption that each voter randomly votes with equal probability for one of the candidates is far from realistic?

The attempt to link mathematical models to questions about elections (and, more generally, to social science) is fascinating and complicated, and a true pioneer in this study was the Marquis de Condorcet, a mathematician and philosopher, a democrat, a human rights advocate, and a feminist who lived in France in the 18th century.

One of Condorcet's findings, often referred to as Condorcet's paradox, is that when there are three candidates running to lead the country – Tzipi, Shelly and

	Type A	Type B	Type C
Shelly	1	2	3
Tzipi	2	3	1
Ayelet	3	1	2

Figure 9:   Condorcet's paradox demonstrated on three prominent Israeli politicians, Tzipi Livni, Shelly Yachimovich, and Ayelet Shaked (above). There are three types of voters, the same number of voters of each type, with a ranking given in the table above.

Ayelet – it is possible that a majority of voters prefer Tzipi to Shelly, another majority of voters prefer Shelly to Ayelet and yet another majority of voters prefer Ayelet to Tzipi. When we try to evaluate the probability of such a phenomenon, the formulas for noise-stability come into play.

Our last puzzle is about the tension between two main themes in mathematics and science as a whole. The first is the pursuit of new avenues and shortcuts for computations and the second is about understanding our own limitations and difficulties.

Finding unexpected paths and shortcuts for computations characterizes many achievements in mathematics and in science. Some examples are the Gaussian elimination method for solving linear systems of equations, Fourier's use of trigonometric functions for solving the heat equation, and Michael Rabin's surprising discovery that randomness can be helpful for computing. An example from physics is Feynman's diagrams used for calculations of subatomic particles.

Of no less importance is the understanding of limitations and impossibilities. The efforts of hundreds of years to find a formula for solving equations of the fifth degree gave way to the understanding that such a formula does not exist, and this was the starting point of modern algebra (and lies at the core of group theory). The centuries-long efforts to prove parallel axioms through other axioms turned out to be impossible, and this understanding gave rise to other types of non-Euclidean geometry. The attempt to prove mathematically that in mathematics itself there are no contradictions turned out to be impossible (Gödel's theorem). One of the important insights of the 20th century is that the computer is not omnipotent, and that some simply formulated problems will never be solved with the aid of computers. And, of course, the impossibility of traveling faster than the speed of light is one of the famous and mind-boggling insights of science. Yehoshua Bar-Hillel wrote a famous paper in the 1960s on the difficulty of computerized translation between languages. He did not point to a fundamental limitation that would never be overcome, but to severe difficulties that many members of the artificial intelligence community of his generation had missed.

We shall now proceed to ask about the possibility of quantum computers.

Puzzle Number 4: Is quantum computing possible?

A quantum computer is a hypothetical physical device that exploits quantum phenomena such as interference and entanglement in order to enhance computing power. The study of quantum computation combines fascinating physics, mathematics, and computer science. In the 1990s, Peter Shor discovered that quantum computers would make it possible to perform certain computational tasks hundreds of orders of magnitude faster than ordinary computers and, in particular, would break most of today's encryption methods. At that time, the first doubts about the model were raised: quantum systems are of a "noisy" and unstable nature. Peter Shor himself found a key to a possible solution to the problem of "noise": quantum error-correcting codes and quantum fault-tolerance. (By the way, one of the directions for the creation of quantum codes is called "topological quantum calculation" and it is based on the homology groups that Poincaré discovered at the beginning of the 20th century and that I mentioned in the first puzzle.)

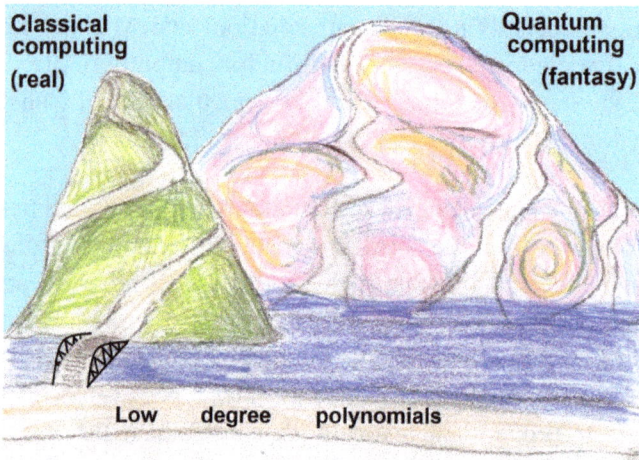

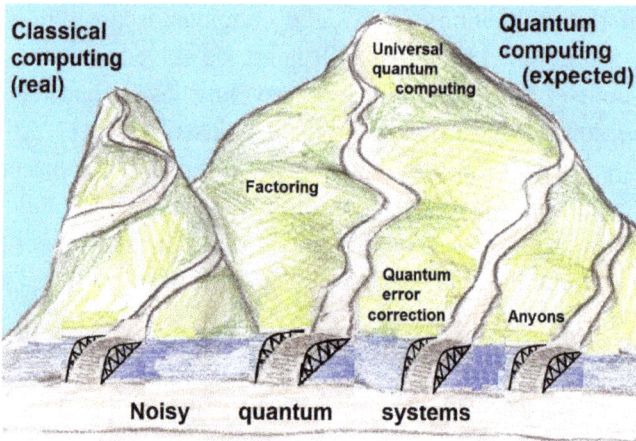

Figure 10:   Are quantum computers possible? My analysis (top); a common view (bottom).

In the mid-1990s, Dorit Aharonov and Michael Ben-Or (and at the same time two other groups) studied the noisy quantum computer model and showed that "noisy" quantum computers still make it possible to perform miracles, as long as engineers succeeded in lowering the noise level below a certain threshold.

One possibility, which reflects widespread opinion, is that the construction of quantum computers is possible, that the remaining challenge is essentially of an engineering nature, and that such computers will be built in the coming decades. Moreover, people expect to build in the next few years quantum codes of the quality required for quantum fault-tolerance, and to demonstrate the concept of "quantum computational supremacy" on quantum computers with 50 computing elements (which are called "qubits"). A second possibility, which expresses my position, is that it will not be possible to construct quantum codes that are required for quantum computation, nor will it be possible to demonstrate quantum computational superiority in other quantum systems. I'll try to explain why.

My research is based on the same model of noise that led researchers in the 1990s to optimism about quantum computation, and it points to the need for different analyses on different scales. My analysis shows that noisy quantum computers in the small scale (a few dozen qubits) express such a primitive computational power that it will not allow the creation of quantum codes that are required as building blocks for quantum computers on a higher scale.

A reader may ask: how does the noisy quantum world allow classical information and classical computation? The difference between classical and quantum information is related to puzzle number 3. Encoding using repetitions and decoding using the majority rule (or related techniques) enable classical information and calculation: the majority method "corrects errors" and is stable to noise, and therefore can be implemented in a very primitive computational system. This enables, on larger scales, classical information and computation.

In summary, the important results from the mid-90s show that once high quality quantum computers are built for roughly 50-300 qubits then it will be possible in principle to use quantum codes to amplify this achievement for building quantum computers with unlimited numbers of qubits. This is correct! The interpretation of these results took for granted that quantum computers with a few tens of qubits are feasible. This, however, is incorrect! A more careful analysis of such quantum computers shows that they can only approach noise-stable (pure) quantum states and those are far too weak for building quantum codes. This prediction about the feasible states for small quantum computers could be tested in the near future as both Google and IBM (and other players as well) are planning to build in a few

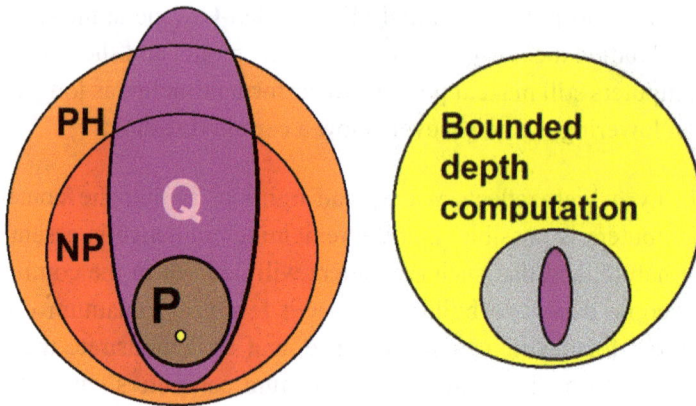

Figure 11:   Quantum computers offer mind-boggling computational superiority (left), but in the small scale, noisy quantum circuits are computationally very weak, unlikely to allow quantum codes needed for quantum computers (right). This figure is a schematic (conjectural) view of the world of computation. Here are more details that are a bit technical. The picture on the left describes some principal classes of computational complexity. The class **P**, represented by a brown circle, refers to "polynomial time" and represents the power of classical computers. **NP**, represented by a red circle, is a larger class of algorithmic tasks that is believed to be intractable for classical computers. (**PH** is an even larger class.) **Q**, represented by the purple ellipse, represents the power of quantum computers. **Q** does not include **NP** but it goes beyond **NP** as well as beyond **PH**. The little yellow circle on the left represents "bounded depth computation" a very primitive form of computing. In the picture on the right, the yellow circle of bounded depth computation is enlarged and the grey ellipse describes an even more primitive class studied by Guy Kindler and myself that has to do with polynomials of small degrees. The purple ellipse represents the power of noisy quantum systems in the small scale.

months (by the end of 2017 or shortly afterwards) quantum computers with 50 qubits.

The quantum computer puzzle is extremely fascinating. Will it be possible to break the limitations of classical computers and speed up the calculation time for certain problems by hundreds of orders of magnitude? In the near term, will it be possible to build quantum codes that will enable stable quantum information that will serve as building blocks for quantum computers? And can quantum computational superiority soon be demonstrated by simple quantum systems? Approximately one and a half billion dollars a year are allocated to dozens of research groups (mainly experimental) for unclassified research on quantum technologies, and much of this is intended to achieve the goals I mentioned. According to my analysis, all these goals are doomed to fail!

Has my alternative analysis of the same model convinced the experts who proposed and investigated it to change their position? Not at all! Other experts argue that errors and noise can in no way be a fundamental obstacle to quantum computing. The intuition of many experimental physicists is also leaning towards optimism. If my position turns out to be correct, one might expect that besides disappointment, understanding the failure of quantum computers will have important implications on the study of quantum systems. Of course, I am very curious to see what it will be like. Richard Feynman famously claimed that if all mathematics disappeared, it would set physics back precisely one week (I never understood what he meant by this). If my analysis of quantum fault-tolerance is correct this will be a nice example of mathematical reasoning that would set an area of physics back two decades! With a bit of luck it will provide us with some useful reflections on theory, computation, and the foundations of quantum physics going back a couple decades further.

In any case, even in mathematics, and especially in places where mathematics meets other scientific fields and real life, it is possible to find interesting and exciting cases of disagreements and debates.

I will conclude by briefly mentioning a puzzle that has recently piqued my interest.

Puzzle Number 5: How to model random RNA trees?

RNA (ribonucleic acid) is an important biological molecule that is built (like other molecules that play a major role in biology) as a chain of smaller molecules called nucleotides. RNA is built from four types of nucleotides represented by the letters G, U, A, and C, and an RNA molecule can be regarded as a sequence of these letters. (This sequence is called the primary structure of an RNA molecule.) An RNA molecule tends to fold on itself as some segments in this long sequence are tied with other segments. This leads to a tree structure as seen in the figure below. (The tree structure is part of the secondary structure of the RNA molecule.) Professor Avinoam Ben-Shaul of the Hebrew University's Institute of Chemistry asked me about mathematical properties of trees formed of random RNA molecules that he and his coauthors had studied. Trees, and random models of trees, are very exciting objects and Avinoam and I are currently exploring this puzzle.

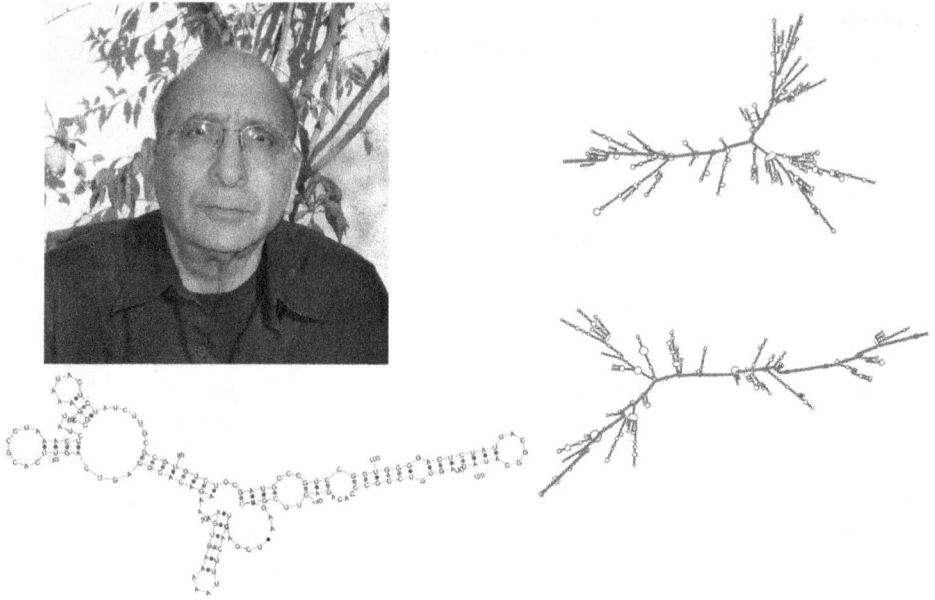

Figure 12:   Avinoam Ben-Shaul with three RNA trees and one lemon tree.

Great Thinkers Praise "Gina Says"

"Gina Says" — The ideal Book to cheer you up in long flights and when you are sick and cannot do other things!

Shmuel Weinberger (August 07): Very much enjoyed the story of Gina's involvement in the blog world — I read it through on my flight back to America and it was a very interesting! The part that resonated most was the advice (I think it came from your father) that every subject is fascinating after you study it deeply.

Avi Wigderson: (August 07) I expected no less from the author of the immortal translation of the classic book "Where is Pluto?"

Itai Benjamini (Feb 08): It was such a joy to listen to your unique voice (music) yet again.

Oded Schramm (Dec 2007): There were some definite enjoyable highlights. (Feb. 08) What about a sequel? I'm really curious what's happening with Gina these days?

Elchanan Mossel (Feb 08): I read it when I was sick and couldn't do other things, and it cheered me up.

Olle Haggstrom (March 08): I found it a real page-turner, and read the entire thing for four straight hours last night! Very interesting stuff, on several levels. And very original, of course. May I ask what is your relation to Gina? You seem to have remarkable insight into her mind...

Ken Binmore (June 08) Dear Gil, I like your book a lot. If you get it published, it could do with some pictures. Best wishes, Ken.

Eva Ilous (Apr 09) I find it very entertaining.

Tselil Schramm (July 09) I have just this moment finished reading the first section of Gina Says, after starting it this morning. I enjoyed it a lot! It was pretty great. I think my favorite part is when Gina gets kicked off, and Gil Kalai posts instead. I laughed out loud! Silly String Wars was also priceless.... (August 09) I finished Gina. I liked it a lot. I liked the story about the imagined conversation with the grad student, does it have basis in reality? I also enjoyed the part about trolls a lot. When I read the story of Miles and Jeffery, I felt like I had heard it before... is this possible?

Greg Kuperberg: (Oct 08) ...So the assertion that spacetime is 10-dimensional comes with the important asterisk that it might only look 10-dimensional when it is near a classical limit in the sense of standard semiclassical perturbation theory. In the realistic limit it looks effectively 4-dimensional, and in more exotic limits it may look 11-dimensional (say). If you do not take a limit at all, then maybe spacetime is some non-geometric algebraic object that doesn't have a dimension.

Scott Aaronson (Nov 08): It is one of the strangest documents I have ever read...

At the beginning, the greatest charm of your manuscript is its promise to bring an "outsider perspective" to the string wars: unlike the partisans on both sides (Motl, Distler, Smolin, Woit...), Gina presents herself as a curious outsider who's just trying to pose questions and understand what's going on. By the end, though, she's become (at least in my eyes) basically another partisan, for the pro-string side... there's lots of food of thought in this unusual manuscript — [despite the above concerns,] I certainly found myself reading till the end.

Clifford Johnson (Nov 08): I find the Gina character rather well done.... I actually really like your idea... it is fresh and amusing. Well done.

Peter Woit (Jan 09): I read through it quickly, amused to relive again some battles of the string wars.... In your fantasy of the future, you mention my book being translated into Czech. Funny, a publishing company there did buy the rights a year or so ago, and I think they will be bringing it out. Sometimes reality and fantasy are indistinguishable in this story...

Motty Perry (Jan 09): I started reading your book on my way to the UK (will be back in 3 days) and was forced to stop when the laptop's battery died. So far (50 pages), it is fascinating!!!; I find it educating and I wait to my next flight to continue reading.

Caterina Calsamiglia (Jan 09) Every once in a while there are interesting thoughts about what the aim of science should be, that are nice...! Physicists and mathematicians are quite special...!

Yuval Peres (Jan 09): I spent several hours reading through it, trying to separate fact from fiction. The statement "Why should I be surprised if I can simply disbelieve." on page 5 is not due to Gina's great uncle "Lena" but rather to my grandmother Malka Heller (right).

Mark Srednicki (June 09): I have to say that it's a bit disappointing to find out that Gina is actually an accomplished mathematician. I remember being impressed by the prescience of some of her comments, back when I believed her to be an interested non-expert, the sort of person the bloggers are all trying to reach.

Janos Pach (April 09): Your book is sensational! I have never seen anything like this. And it partially answers my question that I bothered you with several months ago: what is blogging?

Luca Trevisan (June 09) It was a wonderful idea, and very well executed; but the Comic Sans hurts my eyes.

Marni D Sheppeard (**Kea**) (July 09): Loved the book, thanks! Best of luck with the publishing... This could be a bestseller!... From Gina's sister.

Lubos Motl (July 09): Gina was almost certainly a fictitious nickname of an Israeli mathematical transvestite.

Jacques Distler (June 09): I recall rather enjoying the Gina character at the time. She seemed a curious mix of occasionally penetrating insights, and a sometimes rather befuddling layman's (lack of) understanding of the scientific issues under discussion. ... the book (at least, the excerpt) has a quaintly antiquarian feel. Comic Sans notwithstanding.

Lance Fortnow (June 09): It certainly was a fun read but I finished not sure of the purpose of the book...

John Sidles (June 09): Congratulations Gil ... this book is a very interesting read!

Eytan Sheshinski (July 09) Gil, I am in the middle of the book. It is a fascinating discussion, some tongue and cheek, by smart people on foundational issues of science which we, economists, hardly touch. I will use "why should I be surprised if I can simply disbelieve"!

Jasone Fagone (Oct 15): I stumbled across your book "Gina Says" the other week. It's honestly one of the weirdest and greatest things I've ever seen on the Internet.

www.ingramcontent.com/pod-product-compliance
Lightning Source LLC
Chambersburg PA
CBHW061810210326
41599CB00034B/6954